Benjamin Thompson

Philosophical Transactions

New Experiments upon Gun-Powder

Benjamin Thompson

Philosophical Transactions
New Experiments upon Gun-Powder

ISBN/EAN: 9783337159627

Printed in Europe, USA, Canada, Australia, Japan

Cover: Foto ©berggeist007 / pixelio.de

More available books at **www.hansebooks.com**

PHILOSOPHICAL

TRANSACTIONS.

XV. *New Experiments upon Gun-powder, with occasional Observations and practical Inferences; to which are added, an Account of a new Method of determining the Velocities of all Kinds of Military Projectiles, and the Description of a very accurate Eprouvette for Gun-powder.* By Benjamin Thompson, *Efq, F. R. S.*

Read March 29, 1781.

THESE experiments were undertaken principally with a view to determine the moft advantageous fituation for the vent in fire-arms, and to meafure the velocities of bullets, and the recoil under various circumftances. I had hopes alfo of being able to find out the velocity of the inflammation of gun-powder, and to meafure its force more accurately than had hitherto been done. They were begun in the month of July in the year

1778, at Stoneland Lodge, a country feat of Lord GEORGE GERMAIN's, and I was affifted by the reverend Mr. BALE, rector of Withyham, who lives in the neighbourhood.

The weather proved remarkably favourable for our experiments, being fettled and ferene, fo that the courfe of them was never interrupted for a whole day by rain or by any accident. The mercury in the barometer ftood in general pretty high, and the temperature of the atmofphere was very equal, and moderately warm for the feafon. In order that each experiment might, as nearly as poffible, be under fimilar circumftances, they were all made between the hours of ten in the morning and five in the afternoon : and after each difcharge the piece was wiped out with tow till all the infide of the bore was perfectly clean, and as bright as if it had juft come out of the hands of the maker ; and great care was taken to allow fuch a fpace of time to elapfe between the firings, as might render the heat of the piece nearly the fame in every experiment.

A defcription of the apparatus.

The barrel principally ufed in thefe experiments was made by WOGDON, one of the moft famous gunfmiths in London; and nothing can exceed the accuracy with which it is bored, or the finenefs of the polifh on the infide. It is made of the very beft iron, and, agreeably to Mr. ROBINS's advice, I took care to have it well fortified in every part, that there might be no danger of its burfting. Its weight and dimenfions may be feen in the table of the weight and dimenfions of the apparatus, p. 242.

Fig. 1. Reprefents a longitudinal fection of a part of the barrel, with the apparatus firft made ufe of for fhifting the vent from one part of the chamber to another, or rather for moving
the

the bottom of the chamber further from, or bringing it nearer to, the vent, in order that the fire might be communicated to the powder in different parts of the charge.

a, b, reprefent the lower part of the barrel.

c, is the breech-pin, which is perforated with a hole four-tenths of an inch in diameter, the axis of which coincides with the axis of the bore.

Into this hole the fcrew *h, n,* about four inches in length, is fitted ; to the end of which, *n,* that paffes up into the bore, is fixed a pifton *o, p,* which, by means of collars of oiled leather, is made to fit the bore of the piece very exactly. The end of the pifton *p,* neareft the muzzle, is of brafs, and forms a move-able bottom to the bore, which by turning the fcrew *h, n,* by means of the handle *m,* is brought nearer to, or removed fur-ther from, the fixed vent *v,* by which means the powder is lighted at any affignable diftance from the bottom of the charge.

But the length of the bore being altered by moving the pifton, which occafioned a fmall inaccuracy, and fome incon-venience attending the apparatus, it was laid afide, and ano-ther reprefented by fig. 2. was fubftituted in the room of it.

a, b, is a fection of part of the barrel as before, and *c* is the breech-pin, which being perforated with a fmall hole through its center receives the fcrew *f, g,* which is about two-tenths of an inch in diameter, and four inches long. This fcrew being perforated with a very fmall hole, ferves to convey the fire into the chamber of the piece, and by fcrewing it further up into the bore, or drawing it backwards, the fire is communicated to different parts of the charge.

But this method being found to be not intirely free from inac-curacies and inconveniencies, a third was fubftituted in the

room

room of it, which was found to anfwer much better than either of the preceding.

The end of the bore was now firmly clofed by a folid breech-pin *p*, fig. 3. and three vent holes *m*, *n*, and *o*, were made in the barrel; one of them, *m*, even with the bottom of the bore, and the other two at different diftances from it. Any two of thefe vent holes, as *n* and *o* for inftance, being clofed up by folid fcrews, a perforated ferew, or vent tube *v*, was fcrewed into the third, which ferved to contain the priming, and to convey the fire to the powder lodged in the bore of the piece.

Sometimes a longer vent-tube, reprefented by fig. 4. was made ufe of; which, paffing through the powder in the cham-ber of the piece, communicated the fire immediately to that part of the charge that lay in the axis of the bore.

Another vent-tube alfo was ufed occafionally, which differs in many refpects from both thofe that have been defcribed. It is fo conftructed as to convey the fire to the charge; but, as foon as the powder in the chamber of the piece begins to kindle, and the elaftic fluid to be generated, the vent is firmly clofed by a valve, and no part of the generated fluid is permitted to efcape. This I fhall call the *valve-vent*, and it is reprefented by fig. 5. upon an enlarged fcale, that the parts of it may appear more diftinct.

a, *b*, is a longitudinal fection of a fmall portion of the folid fide of the barrel.

c, *d*, is the vent-tube, which is in all refpects like the fhort vent-tube commonly made ufe of, except only that in this the end of the vent-hole (*c*) which goes into the chamber is en-larged in the form of the wide end of a trumpet or funnel.

To this enlarged aperture the valve, *v*, is accurately fitted, and by means of the fmall ftem or tail, *t*, which is fixed to the

valve,

valve, and which paffes up through the vent-hole, and is connected with the fpring S, the valve is preffed, or rather drawn into its place, and the vent is clofed. The ftem of the valve was at firft made cylindrical; but, in order to make way for the priming to pafs down to the valve, one-half of its fubftance was taken away, as is reprefented in the figure.

When this vent is primed, the fpace between the vent hole and the ftem of the valve is filled with fine-grained powder, and the valve is gently opened by preffing upon the end of the ftem till one or more grains of powder lodge themfelves between the valve and the aperture; which preventing the valve from clofing again, a fmall opening is left for the paffage of the flame into the chamber of the piece: therefore, when the priming is lighted, the fire paffing down the vent, and entering the chamber, inflames the charge, and the fmall grains of powder that were lodged between the valve and the aperture being deftroyed by the flame in its paffage through the vent, the valve immediately clofes, and prevents the efcape of any part of the elaftic fluid generated by the inflammation of the powder in the chamber of the piece. The preffure of this fluid upon the valve affifts the action of the fpring, by which means the valve is more expeditioufly and more effectually clofed.

The valve was very accurately fitted to the aperture by grinding them together with powdered emery, and afterwards polifhing them one upon the other. And it is very certain, that no part of the elaftic fluid made its efcape by this vent; for, upon firing the piece, there was only a fimple flafh from the explofion of the priming, and no ftream of fire was to be feen iffuing from the vent, as is always to be obferved when a common vent is made ufe of, and in all other cafes where this fluid finds a paffage.

6　　　　　　　　　　　　　　　　　　　　　In

In order that every part of the apparatus employed in thefe experiments might be as perfect as poffible, all the more delicate parts of it were executed by Mr. FRASER, mathematical inftrument-maker in Duke's Court, St. Martin's Lane, and, among the reft, all the contrivances juft defcribed relative to the vent.

The velocities of the bullets were determined by means of a pendulum, according to the method invented by Mr. ROBINS.

The pendulum I made ufe of (fig. 6.) is compofed of a circular plate of hammered iron (*a*), 13 inches in diameter, and 0,65 of an inch thick, to which is firmly faftened a bar of iron (*b*, *c*) 56,5 inches in length, 2,6 inches broad, and half an inch in thicknefs, by which it is fufpended by means of two pivots (*d*, *e*) at the end of the bar (*c*), and at right angles to its length. Thefe pivots being very accurately finifhed, and moving on polifhed grooves, which were kept conftantly oiled to leffen the friction, the vibration of the pendulum was very free, as appeared by the great length of time its vibrations continued after it had been put in motion, and was left to itfelf. To the circular plate of the pendulum, targets of circular pieces of wood of different thickneffes were fixed, which in the courfe of the experiments were often fpoiled and replaced : and, in order to mark the weight and dimenfions of the pendulum in each experiment, the pendulums are numbered according to the different targets that were made ufe of; and the weight and dimenfions of each pendulum are fet down in a table at the end of the defcription of the apparatus.

The target of the pendulum N° 1. was made of a circular piece of elm-plank, 3¼ inches thick, and equal in diameter to the iron plate of the pendulum to which it was fixed ; but this target being too thin was very foon ruined.

The

The pendulum N° 2. was furnished with two targets, which were circular pieces of very tough oak-plank, near five inches thick, placed on opposite sides of the plate of the pendulum, and firmly fixed to it by screws, and to each other by iron straps. When one of these targets was ruined, the pendulum was turned about, and the other was made use of. This pendulum lasted from experiment N°9. to experiment N°39. when it was so much shattered as to be rendered unfit for further service.

The pendulum N° 3. was like the pendulum N° 2.; only, instead of oak, elm-plank near seven inches in thicknefs was made use of for the targets. This pendulum served from experiment N° 40. to experiment N° 101. inclusively.

But finding that targets made of planks of the toughest wood were very soon shattered to pieces by the bullets, I composed the pendulum N° 4. in a different manner. Instead of circular pieces of plank, solid cylinders of elm-timber were made use of for the targets, so that the bullets now entered the wood in the direction of its fibres. These cylinders are 13 inches in diameter, and about 5½ inches in length, hooped with iron at both their ends to prevent their splitting, and firmly fastened to the plate of the pendulum, and to each other by four iron straps. This pendulum lasted till the experiments were finished. It is still in being, and appears to be very little the worse for the service it has undergone.

Fig. 7. shews the two ends of the pendulum upon a large scale, together with the hooks or grooves by which it was suspended.

a, *b*, is the bar of the pendulum, which is seen broken off, as there is not room to shew the whole of its length.

c, *d*, are the pivots by which it was suspended.

e, is

e, is the circular plate of the pendulum, to which

f, g, two circular targets, are faftened by fcrews, and by means of the iron ftraps, 1, 2, 3, 4, which are nailed to the edges of the targets.

h, k, are the hooks which ferved inftead of grooves to receive the pivots, *c, d,* of the pendulum.

The hooks were firmly fixed to the horizontal beam R. S. which fupported the whole apparatus by means of three fcrews *m, n, o,* which paffed through three holes in the plate that connects the two hooks. When the hooks were faftened to the beam, the middle fcrew, *n,* was firft put into its place, and the pendulum was allowed to fettle itfelf in a pofition truly perpendicular, after which the grooves were immoveably fixed by means of the fcrews *m, o.*

The chord of the arc, through which the pendulum afcended in each experiment, was meafured by a ribbon, according to the method invented and defcribed by Mr. ROBINS.

The recoil was meafured in the following manner. The barrel was fufpended in an horizontal pofition (and nearly in a line with the center of the target) by two fmall pendulous rods, 64 inches in length, and 25,6 inches afunder; which being parallel to each other, and moving freely upon polifhed pivots about the axes of their fufpenfion, and upon two pair of trunnions that were fixed to the barrel, formed, together with the barrel, a compound pendulum; and from the lengths of the vibrations of this pendulum, the velocity with which the barrel began to recoil, or rather its greateft velocity, was determined.

But in order that the velocity of the recoil might not be too great, fo as to endanger the apparatus when large charges were

made

made ufe of, it was found neceffary to load the barrel with an additional weight of more than 40 lbs. of iron.

This additional weight of iron, which I fhall call the *gun carriage*, as it was fo conftructed as to ferve as a carriage to the barrel, is compofed of a bar of hammered iron 28 inches in length, 2,6 inches broad, and half an inch in thicknefs, which is bent in the middle of its length in fuch a manner, that its two flat fides or ends are parallel to each other, and diftant afunder two inches. In the middle of this bar where it is bent is a hole in the form of an oblong fquare, which, receiving the end of the breech-pin, fupports the lower end or breech of the barrel. The other end of the barrel is fupported and confined in the following manner. A ring or hoop of iron, near half an inch thick, and two inches in diameter, is placed in a vertical pofition between the parallel fides of the bar, and near its two ends, and firmly fixed to them by fcrews. The barrel paffing through the middle of this ring is fupported upon the ends of three fcrews, which paffing through the ring in different parts of its circumference all point towards its center.

The carriage, together with the barrel, was fufpended by the pendulous rods by means of two pair of polifhed trunnions that are fixed to the outfide of the carriage. They are placed in an horizontal line perpendicular to, and paffing through, the axis of the bore.

Fig. 8. reprefents the barrel fixed to the carriage.

a, *b*, *c*, is the bar of iron which forms the carriage feen edge-ways.

2, 2, 4, 4, are the trunnions by which it was fufpended.

d, *e*, is the barrel in its proper place.

p, is the breech-pin, which paffing through a hole in the middle of the bar, *a*, *b*, *c*, fupports the end, *e*, of the barrel; and

n, is the ring that fupports the end, *d*, of the barrel.

Fig.

Fig. 9. reprefents a perpendicular fe&ion through the line 2, 2, fig. 8. and in a line perpendicular to the length of the barrel.

This figure is defigned to fhew the manner in which the muzzle of the piece was fupported and confined in the ring *n*, fig. 8.

a, c, are the two ends of the bar that are feen cut off.

n, is the ring, and

o, p, are the fcrews by which it is faftened to the two parallel fides of the bar, the ends of which form the trunnions 2, 2, fig. 8.

d, is a tranfverfe fe&ion of the barrel, and

r, s, t, are the three fcrews by which the barrel is fupported and confined in the center of the ring.

Fig. 10. is the fame as fig. 9. but upon a larger fcale.

Fig. 11. reprefents the two ends of one of the pendulous rods by which the barrel was fufpended; and fig. 13. fhews the fame feen fideways.

a, b, is the rod which is feen broken off.

c, d, are the pivots by which it was fufpended by a pair of hooks or grooves that were faftened to an horizontal beam, in the fame manner as the pendulum for meafuring the velocities of the bullets was fufpended.

e, f, are the hooks which receive the trunnions that are fixed to the carriage.

The dimenfions of every part of this apparatus may be feen in the table, p. 242.

The chord of the arc through which the barrel afcended in its recoil was meafured by a ribbon, and the lengths of thofe chords, expreffed in inches and decimal parts of an inch, are fet down in the tables. The method of computing the velo-

6

city

city of the recoil from the chord of the arc through which the barrel afcended, is too well known to require an explanation : and it is alfo well known, that the velocities are to each other as the chords of thofe arcs. The lengths of thofe chords, therefore, as they are fet down in the tables, are, in all cafes, as the velocities of the recoil.

The powder made ufe of in thefe experiments was of the beft kind, fuch as is ufed in proving great guns at Woolwich. A cartridge, containing 12 lbs. of this powder, was given to me by the late General DESAGULIERS of the Royal Artillery, and Infpector of Brafs and Iron Ordnance; who alfo, in the politeft manner, offered me every other affiftance in his power towards completing the experiments I had projected, or in making any others I fhould propofe that might be ufeful in the profecution of my inquiries.

This powder was immediately taken out of the cartridge, and put into glafs bottles, which were previoufly made very clean and dry ; and in thefe it was kept carefully fealed up till it was opened for ufe. When it was wanted for the experiments, it was weighed out in a very exact balance, with fo much attention, that there could not poffibly be an error in any inftance greater than one quarter part of a grain. The bottles were never opened but in fine weather, and in a room that was free from damp, and no more charges of powder than were necef- fary for the experiments of the day were weighed out at a time. Each charge was carefully put up in a cartridge of very fine paper, and thefe filled cartridges were kept in a turned wooden box, that was varnifhed on the infide as well as the outfide, to prevent its imbibing moifture from the air.

The paper of which thefe cartridges were made was fo fine and thin, that 1280 fheets of it made no more than an inch in

thicknefs,

thicknefs, and a cartridge capable of containing half an ounce of powder weighed but three quarters of a grain.

The cartridges were formed upon a wooden cylinder, and accurately fitted to the bore of the piece, and the edges of the paper were faftened together with pafte made of flour and water.

When a cartridge was filled, the powder was gently fhaken together, and its mouth was tied up and fecured with a piece of fine thread; and when it was made ufe of it was put intire into the piece, and gently pufhed down into its place with the ramrod, and afterwards it was pricked with a priming-wire thruft through the vent, and the piece was primed; fo that no part of the powder of the charge was loft in the act of loading, as is always the cafe when the powder is put loofe into the barrel: nor was any part of it expended in priming; but the whole quantity was fafely lodged in the bottom of the bore or chamber of the piece, and the bullet was put down immediately upon it, without any wadding either between the cartridge and the bullet, or over the bullet.

The bullets were all caft in the fame mould, and confequently could not vary in their weights above two or three grains at moft, efpecially as I took care to bring the mould to a proper temperature as to heat before I began cafting; and when leather was put about them, or other bullets than thofe of lead were made ufe of, the weight was determined very exactly before they were put into the piece.

The diameter of the bullet was determined by meafurement and alfo by computation from its weight, and the fpecfic gravity of the metal of which it was formed; and both thefe methods gave the fame dimenfions very nearly.

The apparatus was put up for making the experiments in a coach-houfe, which was found very convenient for the purpofe, as the joifts upon which the floor over head was laid afforded a firm and commodious fupport for fufpending the pendulum and the barrel, and the walls and roofs of the building ferved to fcreen the apparatus, which otherwife might have been difcompofed by the wind, and injured by the rain and dews. A pair of very large doors, which formed the whole of one end of the room, were kept conftantly open during the time the experiments were making, in order to preferve the purity of the air within the houfe, which otherwife would have been much injured by the fmoke of the gun-powder ; and that, in all probability, would have had a confiderable effect in leffening the force of the powder, and vitiating the experiments. In order ftill further to guard againft this evil, the barrel was placed as near as poffible to the door, and the pendulum was hung up at the bottom of the room.

Fig. 12. reprefents the apparatus as it was put up for making the experiments.

a, *b*, is the barrel with its carriage, fufpended by the pendulous rods *c*, *d*, and

R, is the ribbon which ferved to meafure the afcending arc of its recoil.

P, is the pendulum, and

r, the ribbon that meafured the arc of its vibration.

The diftance from the mouth of the piece to the pendulum was juft 12 feet.

A table

A table shewing the weights and dimensions of all the principal parts of the apparatus.

Of the barrel.

	Inches.
Length	44,7
Length of the bore from the muzzle to the breech-pin	43,45
Diameter of the bore	0,78
Thickness of metal at the lower vent . .	0,36
Thickness of metal at the muzzle . .	0,1

Weight of the barrel, together with the folid breech-pin, and the vent-fcrews and vent-tube, 6 lbs. 6 oz.

Of the gun carriage.

Length	28,4
Diftance between the two pair of trunnions	25,6
Diameter of each trunnion . .	0,25

Weight 40 lbs. 14 oz.

Of the rods by which the carriage was fufpended.

Length from the axis of fufpenfion, or center of the pivots, to the center of the trunnions of the gun carriage, 64 inches.

Weight of each rod, 1 lb. 4 oz.

Total weight of the barrel and its carriage, together with the allowance that was made for the weight of the rods by which it was fufpended, 48 lbs.

N. B. This was its weight from experiment N° 3. to experiment N° 123. inclufive.

Of

Of the bullet.

Diameter 0,75 of an inch.
Weight in lead 580 grains.

Of the pendulum.

Inches.

Total length of the pendulum from the axis of fufpenfion to the bottom of the circular plate   .   69,5

Diameter of the circular plate to which the targets were
faftened   .   ..   .   13,⁻

Diftance between the fhoulders of the pivots   .   3,8

Diameter of the pivots   .   .   .   ,27

Weight of the iron part of the pendulum 47 lb. 4 oz.

*Of the pendulum with the targets fixed to it, as it was prepared
for making the experiments, and numbered.*

	Total length to the ribbon.	Diftance from the axis of fufpenfion.		Total weight of iron and wood.
		To the center of gravity.	To the center of ofcillation.	
	Inches.	Inches.	Inches.	lbs. oz.
Pendulum N° 1	69,25	50,25	58,45	57 0
—— N° 2	69,5	54,4	59,15	82 4
—— N° 3	——	55,62	60,23	100 12
—— N° 4	——	54.6	59,18	88 4

N. B. The meafure is Englifh feet and inches, and the
weight is avoirdupois.

Having

Having now gone through the defcription of all the principal parts of the apparatus, I fhall proceed to give an account of the experiments. And as it may be fatisfactory to the Society to fee the method of conducting thefe enquiries, as well as the refult of them, I fhall firft give a table of the experiments in the exact order in which they were made, together with my original remarks ; I fhall then make fuch general obfervations as may occur : and afterwards I fhall felect, combine, and compare them, in the manner which beft anfwers the different purpofes to which I fhall apply them.

General table of the experiments.

In the two firſt experiments the barrel was fixed to a carriage (that has not been deſcribed) which, together with the barrel and rods by which it was ſuſpended, weighed only 23½.

Length of the bore of the piece 43,5 inches.

Weight of the bullet 580 grains.

The *pendulum*, N° 1.

Order of the experiments.	The charge of powder.		Vent from the bottom of the charge.	Chord of the aſcending arc of the pendulum.	The bullet ſtruck the target below the axis of the pendulum.	Chord of the arc of the recoil.	Velocity of the bullet.	Remarks.
	Weight.	Height.						
	Grs.	In.	In.	Inches.	Inches.	In.	Ft.in Sec.	
N° 1	208	1,8	0,	13,2	64,5	33,5	1267	Firſt day.
2	. .	. .	,5	14,5	. . .	36,5	1399	

This gun carriage being found to be too light, the other, deſcribed, and repreſented fig. 8. was ſubſtituted in the room of it.

Order of the experiments.	The charge of powder.		Vent from the bottom of the charge.	Chord of the ascending arc of the pendulum.	The bullet struck the target below the axis of the pendulum.	Chord of the arc of the recoil.	Velocity of the bullet.	Remarks.
	Weight.	Height.						
	Grs.	In.	In.	Inches.	Inches.	Inches.	Ft.in fec.	
N° 3	208	1,8	0	12,6	65,	17,8	1213	Second day.
4	..	..	,5	—	—	18,5	—	The pendulum gave way.
5	..	..	0	—	—	38,68	—	4 bullets were fired at once.
6	..	..	,5	—	—	38,48	—	Ditto.
7	..	..	0	—	—	6,1	—	Without any bullet.
8	416	3,6	..	—	—	16,5	—	Ditto.
9	208	1,8	0	8,5	65,	17,69	1281	Pen. N°2; very fair; 3d day.
10	104	,9	..	5,2	65,25	10,18	782	
11	310	2,7	0	9,6	64,6	24,69	1459	⎱
12	..	..	1,22	10,1	65,	24,95	1527	The powder was lighted
13	..	..	2,65	11,85	64,75	24,9	1801	by the long vent-tube
14	..	..	..	10,9	65,25	...	1646	(fig. 4.).
15	330	2,9	2,65	10,9	61,5	26,2	1748	
16	..	..	..	13,25	63,5	...	2060	
17	330	1,7	2,65	...	...	12,7	..	The barrel very much heated.
18	..	2,9	0	10,4	63,5	26,3	1619	
19	..	..	..	...	63,	26,4	1633	
20	165	1,45	0	6,8	62,2	14,73	1084	
21	..	..	..	6,85	...	14,2	1093	
22	..	..	1,32	6,7	...	14,8	1071	
23	..	..	..	6 3	60,6	14,58	1035	⎧ The short vent-tube (*v*,
24	..	..	..	7,5	61,5	14,68	1142	fig. 3.) was made use of.

In order to determine how much of the force of the powder was loſt by windage and by the vent, oiled leather was faſtened round the bullet, ſo that it now accurately fitted the bore of the piece; and in the five experiments, from N° 35. to N° 39. incluſive, the valve-vent was made uſe of.

Weight of the bullet, together with the leather in which it was enveloped, 603 grains.

Order

Order of the experiments.	The charge of powder.		Vent from the bottom of the charge.	Chord of the ascending arc of the pendulum.	The bullet struck the target below the axis of the pendulum.	Chord of the arc of the recoil.	Velocity of the bullet.	Remarks.
	Weight.	Height.						
	Grs.	In.	In.	Inches.	Inches.	Inches.	Ft.in sec.	
N°								
25	165	1,45	o	6,8	65,	14,95	1004	Fourth day.
26	..	..	..	7,8	...	15,6	1153	
27	..	..	..	8,05	...	16,15	1192	
28	330	2,9	..	10,2	63,	26,	1559	
29	..	..	2,6	...	64,	28,1	1536	
30	165	3,2	..	5,9	62,4	13,2	914	
31	..	1,45	1,3	6,65	62,6	15,15	1027	

Finding that the blaſt of the powder always reached as far as the pendulum, when large charges were made uſe of, and ſuſpecting that this circumſtance, together with the impulſe of the unfired grains, might in a great meaſure occaſion the apparent irregularity in the velocities of the bullets; to remedy theſe inconveniences, a large ſheet of paper of a moderate thickneſs was ſtretched upon a ſquare frame of wood, and interpoſed as a ſcreen before the pendulum at the diſtance of two feet from the ſurface of the target.

Two reaſons conſpired to induce me to prefer this method of preventing the impulſe of the flame upon the pendulum to the obvious one of removing the pendulum further from the mouth of the piece; the firſt was, that I was unwilling to increaſe the diſtance between the barrel and the pendulum, leſt the reſiſtance of the air might affect the velocities of the bullets; and the ſecond, which I confeſs did not operate leſs ſtrongly than the firſt, was, that the length of the houſe did not admit of a greater

2 L 2   diſtance,

diftance, and I was unwilling to expofe any part of the apparatus in the open air.

But the fcreen was found to anfwer perfectly well the purpofe for which it was defigned, and it was continued during the
remainder of the experiments, the paper being replaced every
third or fourth experiment.

The experiments continued.

Order of the experiments.	The charge of powder.		Vent from the bottom of the charge.	Chord ot the afcending arc of the pendulum.	Bullet ftruck the target below the axis of the pendulum.	Chord of the arc of the recoil.	Velocity of the bullet.	Remarks.
	Weight.	Height.						
N°.	Grs.	In.	In.	Inches.	Inches.	Inches.	Ft.in fec.	
32	165	1,45	0	5,45	63,	15,45	839	Not leathered; weight of the bullet and wad 603 grs. In exp. N° 32. no lefs than 40 large grs. of unfired powder were driven through the fcreen.
33	..	..	..	...	...	12,65	839	
34	..	..	..	7,9	...	15,45	1217	In thefe 6 experiments the bullets were leathered, and the powder was lighted by the valve-vent.
35	..	..	..	7,	60,25	15,25	1129	
36	..	..	..	7,4	62,	16,3	1161	
37	..	..	1,3	8,	61,	17,9	1277	
38	290	2,6	2,6	9,	58,6	23,5	1497	
39	..	..	..	...	...	24.8	..	The pend. N° 2. ruined.

The bullets were now put naked into the piece, and the
powder was lighted by the fhort vent-tube (*v*, fig. 3) and fome
little improvement was made in the fteel edges between which
the ribbons paffed that ferved to meafure the afcending arcs of
the pendulum and of the recoil, by which means the friction
was leffened, and the ribbon was prevented from twifting or
entangling itfelf as it was drawn out.

3   *Apparatu*

Apparatus.

The barrel with its carriage as before.
The pendulum, N° 3. and
Leaden bullets, weighing 580 grains each.

Order of the experiments.	The charge of powder. Weight.	Height.	Vent from the bottom of the charge.	Chord of the afcending arc of the pendulum.	Bullet ftruck the target below the axis of the pendulum.	Chord of the arc of the recoil.	Velocity of the bullet.	Remarks.
N°	Grs.	In.	In.	Inches.	Inches.	Inches.	Ft.in fec.	
40	218	1,9	0	6,45	64,6	18,	1236	5th day; medium velocity in thefe experiments and N° 47. 1225.
41	..	..	..	6,31	65,3	17,71	1197	
42	..	..	..	6,45	65,	17,91	1230	
43	..	..	1,3	6,5	64,6	18,3	1248	Medium velocity 1276
44	..	..	..	6,75	64,5	18,35	1299	
45	..	..	..	6,6	64,9	...	1265	
46	..	..	..	6,4	61,6	...	1293	
47	..	..	0,	6,3	62,	18,1	1266	
48	290	2,6	0	7,2	63,5	22,58	1414	Medium velocity 1427.
49	..	..	..	7,4	...	22,92	1455	
50	..	..	..	7,3	64,6	22,38	1412	
51	290	2,6	1,3	7,4	63,	23,21	1476	Medium velocity 1493.
52	..	..	..	7,6	64,	23,76	1520	
53	..	..	..	7,25	61,	23,6	1483	
54	..	..	2,6	7,5	62,3	...	1502	Medium velocity 1460.
55	..	..	..	7,4	64,	23,26	1450	
56	..	..	..	7,1	62,2	...	1433	
57	..	..	..	7,4	64,	23.56	1454	
58	..	..	..	1,31		11,12	—	In thefe 4 experiments the piece was fired with powder alone, and the fcreen was taken away from before the pendulum.
59	..	..	..	1,2		11,62	—	
60	..	..	0	1,16		9,62	—	
61	..	..	1,3	0,6		11,33	—	

Order

Order of the experiments	The charge of powder.		Vent from the bottom of the charge.	Chord of the ascending arc of the pendulum.	Bullet struck the target below the axis of the pendulum.	Chord of the arc of the recoil.	Velocity of the bullet.	Remarks.
	Weight.	Height.						
	Grs.	In.	In.	Inches.	Inches.	Inches.	Ft.in fec.	
Nᵒ								
62	330	2,9	1,3	8,	63,	26,4	1599	6th day; medium velocity 1625.
63	..	..	..	8,5	65,	...	1652	
64	..	..	2,6	7,2	59,5	25,3	1562	Medium velocity 1528.
65	..	..	..	7,7	65,	...	1495	
66	..	..	0	8,4	...	26,35	1633	Medium velocity 1594.
67	..	..	..	8,	...	25,8	1556	
68	218	1,9	0	6,82	64,	19,56	1349	The powder was rammed very hard.
69	..	..	..	6,6	64,6	18,2	1294	Ditto much harder.
70	..	..	..	6,85	...	19,12	1345	Ditto as hard as in N° 68.
71	..	..	1,3	5,5	...	16,33	1080	Ditto, ditto.
72	..	..	0	——	——	8,72	—	Government powder, no bullet.
73	..	..	..	——	——	8,44	—	Best double battle powder.
74	..	..	1,3	——	——	8,47	—	Government powder.
75	..	..	..	——	——	9,3	—	Double battle powder.

The following experiments N° 78, 79, 80, and 81. were made in hopes of being able to difcover a method of adding to the force of gun-powder. Twenty grains of the fubftances mentioned in the remarks upon each experiment were intimately mixed with the powder of the charge. In the experiment N° 82. a large wad of tow, well foaked in etherial fpirit of turpentine, was put into the piece immediately upon the bullet: and in the experiment N° 83. a wad, foaked in alkohol, was put into the piece in like manner.

Order

Order of the experiments.	The charge of powder.		Vent from the bottom of the charge.	Chord of the ascending arc of the pendulum.	Bullet struck the target below the axis of the pendulum.	Chord of the arc of the recoil.	Velocity of the bullet.	Remarks.
	Weight.	Height.						
N°	Grs.	In.	In.	Inches.	Inches.	Inches.	Ft. in fec.	
76	145	1,3	0	5,3	65,	13,25	1037	⎫ 7th day; medium velo-
77	..	..	..	...	64,6	13,25	1044	⎭ city 1040.
78	..	..	..	3,2	...	8 92		20 grs. beſt alkaline falt of tartar.
79	..	..	..	4,35	...	11,68		20 grs. æthiops mineral.
80	..	..	..	3,3	63,6	9,83		20 grs. fal ammon.
81	..	..	..	4,2	63,4	11,45		20 grs. fine brafs duſt.
82	..	..	..	—	—	15,25	—	⎧ The ſcrews which held the hooks ⎪ by which the pendulum was ſuf- ⎨ pended gave way, and the pen- ⎩ dulum came down.
83	..	..	..	—	—	14,35	—	

In the nine following experiments, *viz.* from N° 84. to N° 92. inclufive, the valve-vent was made ufe of, and the bullets were made to fit the bore of the piece very exactly by means of oiled leather, which was fo firmly faftened about them that in each experiment it entered the target with the bullet.

The bullet made ufe of in experiment N° 85. was of wood.

Thofe ufed in the experiments N° 86. and N° 87. were formed in the following manner; a fmall bullet was caft of plaifter of Paris, which being thoroughly dried, and well heated at the fire, was fixed in the center of the mould that ferved for cafting all the leaden bullets made ufe of in thefe experiments; and melted lead being poured into this mould, the cavity that furrounded the fmall plaifter bullet was intirely filled up, and a bullet was produced, which to the eye had every appearance of folidity, but was as much lighter than a folid leaden bullet of

the

the fame diameter as the plaifter bullet was lighter than a leaden bullet of the fame fize.

In the experiments N° 88. and N° 89. folid leaden bullets were made ufe of. In the experiment N° 90. two bullets were difcharged at once ; in the experiment N° 91. three ; and in the experiment N° 92. four were ufed.

In each of thefe experiments a frefh fheet of paper was made ufe of as a fcreen to the pendulum, in order that the velocities of the bullets might be meafured more accurately ; and alfo, that the quantity of unfired powder might be eftimated with greater precifion.

Order of the experiments.	The charge of powder.		Vent from the bottom of the charge.	Weight of the bullet.	Chord of the afcending arc of the pendulum.	Bullet ftruck the target below the axis of the pendulum.	Chord of the arc of recoil.	Velocity of the bullet.	Remarks.
	Weight.	Height.							
	Grs.	In.	In.	Grs.	Inches.	Inches.	Inches.	Ft.in fec.	
N°									
84	145	1,3	0	—	—	—	4,5	—	8th day ; in each of thefe 4 experiments from 50 to 70 granulæ or particles of unfired powder were driven through the fcreen.
85	..	..	..	90	1,33	62,2	7,16	1763	
86	..	..	..	251	2,82	63,2	9,62	1317	
87	..	..	..	354	3,32	61,2	11,3	1136	
88	..	..	..	600	6,5	65,4	15,22	1229	Very few unfired grains of powder ftruck the fcreen.
89	..	..	..	603	6,3	64,6	15,13	1229	
90	..	..	..	1184	10,12	65,	21,92	978	There were no marks of any unfired powder having reached the fcreen.
91	..	..	..	1754	13,65	63,4	27,18	916	
92	..	..	..	2352	16,55	63.3	32,25	833	

In the feven following experiments the piece was fired with
powder only.

Order of the experiments.	The charge of powder.		Vent from the bottom of the charge.	Chord of the afcending arc of the pendulum.	Chord of the arc of the recoil.	Remarks.
	Weight	Height.				
	Grs.	In.	In.	Inches.	Inches.	
N°						
93	145	1,3	0	——	4,3	
94	165	1,45	. .	——	5,5	
95	. .	. .	. .	——	5,6	
96	290	2,6	. .	——	11,70	
97	437½	3,9	. .	1,68	17,5	The fcreen was taken away.
98	. .	. .	. .	6,7	15,88	{ The whole furface of the target was befpattered with unfired grains of powder.
99	. .	. .	. .	——	17,9	The pendulum was not obferved.

In the following experiments N° 100. and N° 101. the bullets were not put down into the bore, but were fupported by
three wires, which being faftened to the end of the barrel pro-
je&ed beyond it, and confined the bullet in fuch a fituation that
its center was in a line with the axis of the bore, and its hinder
part was one-twentieth of an inch without or beyond the
mouth of the piece.

In experiment N° 102. the bullet was juft ftuck into the barrel in fuch a manner that near one-half of it was without the
bore.

Order of the experiments.	The charge of powder.		Vent from the bottom of the charge.	Chord of the afcending arc of the pendulum.	Bullet ftruck the target below the axis of the pendulum.	Chord of the arc of the recoil.	Velocity of the bullet.	Remarks.
	Weight.	Height.						
	Grs.	In.	In.	Inches.	Inches.	Inches.	Ft.in fec.	
Nᵒ 100	165	1,45	0	,65	60,5	4,9	138	In each of thefe experim. near $\frac{1}{50}$th part of the fub-
101	..	..	..	,43	uncertain.	4,8	92	ftance of the bullet was
102	..	..	..	,86	63,	5,6	180	melted and blown away by the impulfe of the flame.

All that part of the bullet which lay towards the bore of the piece appeared to be quite flat from the lofs of fubftance it had fuftained; and its furface was full of fmall indents, which probably were occafioned by the unfired grains of powder that impinged againft it.

The following experiments were made with the pendulum N° 4. The rest of the apparatus as before.

Order of the experiments	The charge of powder.		Vent from the bottom of the charge.	Chord of the ascending arc of the pendulum.	Bullet struck the target below the axis of the pendulum.	Chord of the arc of the recoil.	Velocity of the bullet.	Remarks.
	Weight.	Height.						
	In.	In.	In.	Inches.	Inches.	Inches.	Ft.in fec.	
N°								9th day.
103	104	,9	0	4,51	65,	10,6	732	About 40 grains of powder were driven through the screen.
104	145	1,3	..	5,4	...	12,92	877	About 40 unfired grs. of powder. Medium velocity 894.
105	..	..	..	5,6	...	13,28	910	40 unfired grains.
106	..	1,14	..	6,18	65,8	14,3	990	Double proof battle powder; no unfired grains.
107	218	1,8	..	8,48	65,	19,68	1380	Ditto, ditto.
108	290	2,6	..	9,45	65,6	23,9	1526	Government powder; bullet leathered; weight 602 grains.
109	..	..	..	8,73	65,2	22,8	1419	Bullet naked; very few unfired grains.
110	..	..	..	9,3	65,6	23,4	1460	Medium velocity 1444.
111	..	..	..	...	...	...	1462	
112	..	..	..	8,85	65,5	22,94	1436	
113	..	..	2,6	8,65	64,	23,7	1438	Medium velocity 1413.
114	..	..	..	8,5	63,6	24,1	1423	
115	..	..	..	8,4	65,	23,8	1378	
116	..	2,28	..	9,15	64,	24,6	1525	Double proof battle powder.
117	437½	3,9	..	10,56	64,9	33,	1738	Gov. pow. No unfired grains thro' the screen.
118	..	..	..	11,	64,5	33,3	1824	Medium velocity 1764.
119	..	..	..	10,5	65,	33,6	1729	
120	..	..	2,6	10,35	...	32,5	1706	Medium velocity 1751.
121	..	..	..	10,65	...	33,2	1757	
122	..	..	..	10,6	63,6	32,9	1789	
123	..	..	0	---	---	17,9	—	Without any bullet.

Of

Of the method made use of for computing the velocities of the bullets.

As the method of computing the velocity of a bullet from the arc of the vibration of a pendulum into which it is fired is fo well known, I fhall not enlarge upon it in this place, but fhall juft give the theorems that have been propofed by different authors, and fhall refer thofe who wifh to fee more on the fubject to Mr. ROBINS's New Principles of Gunnery; to Profeffor EULER's Obfervations upon Mr. ROBINS's Book; and, laftly, to Dr. HUTTON's Paper on the initial Velocities of Cannon Balls, which is publifhed in the Tranfactions of the Society for the year 1778.

If a denote the length from the axis of the pendulum to the ribbon which meafures the chord of the arc of its vibration;

g, the diftance of the center of gravity below the axis;

f, the diftance of the center of ofcillation;

b, the diftance of the point ftruck by the bullet;

c, the chord of the afcending arc of the pendulum;

P, the weight of the pendulum;

b, the weight of the bullet, and

v, the original velocity of the bullet;

$$v = \frac{c}{a} \times \frac{\overline{\mathrm{P}g}}{bb} + \frac{\overline{b}}{f} \times \frac{f}{\sqrt{2b}},$$ is a theorem for finding the velocity upon Mr. ROBINS's principles.

$$*v = \frac{c}{a} \times \frac{\overline{\mathrm{P}g}}{bb} + \frac{\overline{f+b}}{2f} \times \sqrt{\frac{f}{2}},$$ is the theorem propofed by Profeffor EULER, who has corrected a fmall error in Mr. ROBINS's method; and

---

* Put the rational part $\frac{c}{a} \times \frac{\overline{\mathrm{P}g}}{bb} + \frac{\overline{f+b}}{2f} = n$, and exprefs f in the thoufandth parts of a Rhynland foot; then the velocity with which the ball ftrikes the pendulum will be $= \frac{n}{4}\sqrt{\frac{f}{2}}$ Rhynland feet in a fecond.

$v = 5,672 \, cg \sqrt{f} \times \frac{P+b}{bba}$ is Dr. HUTTON's theorem, which is
fufficiently accurate, and far more fimple and expeditious than
either of the preceding. It is to be remembered, that g, h,
and c, may be expreffed in any meafure; but f muft be Englifh
feet, and v will be the velocity of the bullet in Englifh feet in
a fecond.

The velocities of the bullets in moft of the foregoing experiments were firft computed by EULER's method, as I had not
then feen Dr. HUTTON's paper; but in going over the calcula-
tions a fecond time, I made ufe of Dr. HUTTON's theorem.
Both thefe methods gave the fame velocity very nearly, but the
Doctor's method is by much the eafieft in practice.

In thefe computations care was taken to make a proper allowance for the bullets that were lodged in the pendulum, and alfo
for the velocity loft by the bullet in paffing through the fcreen.

The corrections neceffary on account of the bullets lodged in
the pendulum were made in the following manner.

b was continually added to the value of P,

$\frac{b-c}{P} \times b$ . . . . to the value of g, and

$\frac{f-b}{P} \times b$ . . . . to the value of f.

Of the fpaces occupied by the different charges of powder.

The heights of the charges of powder, or the lengths of
the fpaces which they occupied in the bore, were determined
by meafurement; and in order that this might be done with
greater accuracy, inches and tenths of inches were marked
upon the ram-rod, and the charge was gently forced down till it
occupied the fame fpace in each experiment.

The following table fhews the heights of the charges as they
were determined by meafurement, and alfo their heights computed.

puted from the diameter of the bore of the piece, and the fpecific gravity of the powder that was made ufe of.

N. B. By an experiment I fhall give an account of hereafter, I found the fpecific gravity of this powder fhaken well together to be to that of rain water as 0,937 is to 1,000.

Weight of the powder.	Height of the charge.	
	Meafured.	Computed.
Grs.	Inches.	Inches.
104	,9	0,8957
145	1,3	1,2490
165	1,45	1,4211
208	1,8	1,7914
218	1,9	1,8775
290	2,6	2,4980
310	2,7	2,6700
330	2,9	2,8422
416	3,6	3,5828
437½	3,9	3,7680

In the experiment N° 30. the powder was put into a cartridge fo much fmaller than the bore of the piece, that the charge, inftead of occupying 1,45 inches, extended 3,2 inches. By this difpofition of the powder, its action upon the bullet appears to have been very much diminifhed.

Of the effect that the heat which pieces acquire in firing produces upon the force of powder.

It is very probable, that the excefs of the velocity of the bullet in the fecond experiment over that of the firft was occafioned more by the heat the barrel had acquired in the firft experiment than by the pofition of the vent, or any other circumftance; for I have fince found, upon repeated trials, that the force of any given charge of powder is confiderably greater when it is fired in a piece that has been previoufly heated by firing, or by any other means, than when the piece has not been heated. Every body that is acquainted with artillery knows, that the recoil of great guns is much more violent after the fecond or third difcharge than it is at firft; and on fhip-board,

where

where it is neceffary to attend to the recoil of the guns, in
order to prevent very dangerous accidents that might be occafioned by it, the conftant practice has been in our navy, and, I
believe, on board the fhips of all other nations, to leffen the
quantity of powder after the firft four or five rounds : our 32
pounders, for inftance, are commonly fired with 14 lbs. of
powder at the beginning of an action, but the charge is very
foon reduced to 11 lbs. and afterwards to 9 lbs. and the filled
cartridges are prepared accordingly.

By the recoil it fhould feem, that the powder exerted a greater
force alfo in the fourth experiment, being the fecond upon the
fecond day, than it did upon the third, or the firft upon that
day; but the pendulum giving way, it was not poffible to compare the velocities of the bullets in the manner we did in the
two experiments mentioned above.

This augmentation of the force of powder, when it is fired
in a piece that is warm, may be accounted for in the following
manner.   There is no fubftance we are acquainted with that
does not require to be heated before it will burn ; even gun-
powder is not inflammable when it is cold.   Great numbers of
fparks or red-hot particles from the flint and fteel are frequently
feen to light upon the priming of a mufket, without fetting
fire to the powder, and grains of powder may be made to pafs
through the flame of a candle without taking the fire; and
what is ftill more extraordinary, if large grains of powder are
let fall from the height of two or three feet upon a red-hot
plate of iron, laid at an angle of about 45° with the plane of
the horizon, they will rebound intire without being burnt, or
in the leaft altered, by the experiment.   In all thefe cafes the
fire is too feeble, or the duration of its action is not fufficiently
long;

long to heat the powder to that degree which is neceffary in order to its being rendered inflammable.

Now as gun-powder, as well as all other bodies, acquires heat by degrees, and as fome fpace of time is taken up in this as well as in all other operations, it follows, that powder, which has been warmed by being put into a piece made hot by repeated firing, is much nearer that ftate in which it will burn, or, I may fay, is more inflammable than powder which is cold; confequently, more of it will take fire in a given fhort fpace of time, and its action upon the bullet and upon the gun will of courfe be greater.

The heat of the piece will alfo ferve to dry the air in the bore, and to clear the infide of the gun of the moifture that collects there when it has not been fired for fome time, and thefe circumftances doubtlefs contribute fomething to the quicknefs of the inflammation of the powder, and confequently to its force.

As it takes a longer time to heat a large body than a fmall one, it follows, that meal-powder is more inflammable than that which is grained; and the fmaller the particles are, the quicker they will take fire. The failors bruife the priming after they have put it to their guns, as they find it very difficult, without this precaution, to fire them off with a match: and if thofe who are fond of fporting would make ufe of a fimilar artifice, and prime their pieces with meal-powder, they would mifs fire lefs often, the fprings of the lock might be made more tender, and its fize confiderably reduced without any rifque, and the violence of the blow of the flint and fteel in ftriking fire being leffened, the piece might be fired with greater precifion.

3   Concluding

Concluding from the refult of the four experiments mentioned above, as well as from the reafons juft cited, that the
temperature of the piece has a confiderable effect upon the force
of the powder, I afterwards took care to bring the barrel to a
proper degree of heat, by firing it once or oftener with powder
each time I recommenced the experiments after the piece had
been left to cool.

Of the manner in which pieces acquire heat in firing.

I was much furprifed upon taking hold of the barrel immediately after the experiment N° 17. when it was fired with 330
grains of powder without any bullet, to find it fo very hot that
I could fcarcely bear it in my hand, evidently much hotter than
I had ever obferved it before, notwithftanding the fame charge
of powder had been made ufe of in the two preceding experiments, and in both thefe experiments the piece was loaded
with a bullet, which one would naturally imagine, by confining
the flame, and prolonging the time of its action, would heat
the barrel much more than when it was fired with powder
alone.

I was convinced that I could not be miftaken in the fact, for
it had been my conftant practice to take hold of the piece to
wipe it out as foon as an experiment was finifhed, and I never
before had found any inconvenience from the heat in holding it.
But in order to put the matter beyond all doubt, after letting
the barrel cool down to the proper temperature, I repeated the
experiment twice with the fame charge of powder and a bullet;
and in both thefe trials (experiments N° 18. and N° 19.) the
heat of the piece was evidently much lefs than what it-was in
the experiment above mentioned (N° 17.).

I now regretted exceedingly the lofs of a fmall pocket ther-
mometer, which I had provided on purpofe to meafure the
heat of the barrel, but it was accidentally broken by a fall the
day before I began my experiments; and being fo far from
London, I had it not in my power to procure another : I was
therefore ob'iged to content myfelf with determining the heat
of the piece as well as I could by the touch.

Being much ftruck with this accidental difcovery of the great
degree of heat that pieces acquire when they are fired with
powder without any bullet, and being defirous of finding out
whether it is a circumftance that obtains univerfally, I was
very attentive to the heat of the barrel after each of the fucceeding experiments ; and I conftantly found the heat fenfibly
greater when the piece was fired with powder only, than when
the fame charge was made to impel one or more bullets.

Though the refult of thefe experiments was totally unex-
pected, and even contrary to what I fhould have foretold if I
had been afked an opinion upon the fubject previous to making
them ; yet, after mature confideration, I am now convinced,
that it is what ought to happen, and that it may be accounted
for very well upon principles that are clearly admiffible.

It is certain, that a very fmall part only of the heat that a
piece of ordnance acquires in being fired is communicated to it
by the flame of the powder; for the time of its action is fo
fhort (not being, perhaps, in general longer than about $\frac{1}{100}$th
or $\frac{1}{130}$th part of a fecond) that if its heat, inftead of being 4
times, as Mr. ROBINS fuppofes, was 400 times hotter than red-
hot iron, it could not fenfibly warm fo large a body of metal
as goes to form one of our large pieces of cannon. And be-
fides, if the heat of the flame was fufficiently intenfe to produce fo great an effect in fo fhort a time, it would certainly be

4   fufficient

fufficient to burn up all inflammable bodies that it came near, and to melt the fhot that it furrounded and impelled, efpecially when they were fmall, and were compofed of lead or any other foft metal; but, on the contrary, we frequently fee the fineft paper come out of the mouth of a piece uninflamed, after it has fuftained the action of the fire through the whole length of the bore, and the fmalleft lead fhot is difcharged without being melted.

But it may be objected here, that bullets are always found to be very hot if they are taken up immediately after they come out of a gun; and that this circumftance is a proof of the intenfity of the heat of the flame of powder, and of its great power of communicating heat to the denfeft bodies. But to this I anfwer, I have always obferved the fame thing of bullets difcharged from wind-guns and crofs-bows, efpecially when they have impinged againft any hard body, and are much flattened; and bullets from mufkets are always found to be hotter in proportion to the hardnefs of the body againft which they are fired. If a mufket ball is fired into any very foft body, as (for inftance) into water, it will not be found to be fenfibly warmed; but if it is fired againft a thick plate of iron, or any other body that it cannot penetrate, the bullet will be demolifhed by the blow, and the pieces of it that are difperfed about will be found to be in a ftate very little fhort of fufion, as I have often found by experience. It is not by the flame therefore that bullets are heated, but by percuffion. They may, indeed, receive fome fmall degree of warmth from the flame, and ftill more perhaps by friction againft the fides of the bore, but it is in ftriking againft hard bodies, and from the refiftance they meet with in penetrating thofe that are fofter, that they acquire by far

2 N 2   the

the greater part of the heat we find in them as foon as they come to be at reft, after having been difcharged from a gun.

There is another circumftance that may poffibly be brought as an objection to this opinion, and that is the running of the metal in brafs guns upon repeatedly firing them, by which means the vent is often fo far enlarged as to render the piece intirely ufelefs. But this, I think, proves nothing but that brafs is very eafily corroded, and deftroyed by the flame of gun-powder; for it cannot be fuppofed, that in thefe cafes the metal is ever fairly melted. The vent of a mufket is very foon enlarged by firing, and after a long courfe of fervice it is found neceffary to ftop it up with a folid fcrew, through the center of which a new vent is made of the proper dimenfions. This operation is called bufhing, or rather bouching the piece; but in all the better kind of fowling-pieces the vent is lined, or bouched, with gold, and they are found to ftand fire for any length of time without receiving the leaft injury. But every body knows that gold will run with a lefs heat than is required to melt iron: but gold is not corroded either by the fpirit of nitre, or the acid fpirit that is generated from fulphur, whereas iron is very eafily deftroyed by either; and that I take to be the only reafon why a vent that is lined with gold is fo much more durable than one that is made in iron. But it feems, that iron is more durable than brafs; and perhaps fteel, or fome other cheap metal, may be found that will fupply the place of gold, and by that means the great expence that attends bouching pieces with that precious metal may be fpared, and this improvement may be introduced into common ufe.

This leads us to a very eafy and effectual remedy for that defect fo long complained of in all kinds of brafs ordnance, *the running of the vent*; for if thefe pieces were bouched with

iron,

iron, there is no doubt but they would ſtand fire as well as iron guns; and if ſteel, or any other metal, either ſimple or compounded, ſhould upon trial be found to anſwer for that purpoſe better than iron, it might be uſed inſtead of it; and even if gold was made uſe of for lining the vent, I imagine it might be done in ſuch a manner as that the expence would not be very conſiderable, at the ſame time that the thickneſs of the gold ſhould be ſufficient to withſtand the force of the flame for a very great length of time.

But to return to the heat acquired by guns in firing. It being pretty evident that it is not all communicated by the flame, there is but one other cauſe to which it can be attributed, and that is the motion and friction of the internal parts of the metal among themſelves, occaſioned by the ſudden and violent effort of the powder upon the inſide of the bore, and to this cauſe I imagine the heat is principally if not almoſt intirely owing. It is well known, that a very great degree of heat may be generated in any hard and denſe body in a ſhort ſpace of time by friction, and in a ſtill ſhorter time by colliſion. " For if two denſe hard elaſtic bodies be ſtruck againſt " each other with great force and velocity, all the parts of " ſuch bodies will every moment be cloſely compreſſed, and " being rigid will re-act with equal force. Hence a quick and " powerful contraction and expanſion will ariſe in every part, " reſembling that ſwift kind of vibrations obſerved in ſtretched " ſtrings; how great theſe vibrations are may be learnt from " the inſtance of a bell, when ſtruck with a ſingle blow, by " which the whole bulk, however vaſt, will for a long time " expand, and contract itſelf in infinite ellipſes. And when " the attrition above deſcribed is produced, with what force and " velocity are all the particles of the rubbed body compreſſed,

2 " ſhaken,

" fhaken, and loofened to their very intimate fubftance*?" And
in proportion to the fwiftnefs of this vibration, and the violence
of the attrition and friction, will be the heat that is produced.

A piece of iron that would fuftain the preffure of any weight,
however large, without being warmed, may be made quite hot
by the blow of a hammer; and even foft and un-elaftic bodies
may be warmed by percuffion, provided the velocity with which
their parts are made to give way to the blow is fufficiently rapid.
If a leaden bullet is laid upon an anvil, or any other hard body,
and in that fituation it is ftruck with a fmart blow of a hammer, it will be found to be much heated; but the fame bullet in
the fame fituation may be much more flattened by preffure, or
by the ftroke of a very heavy body moving with a fmall velocity, without being fenfibly warmed.

To generate heat therefore the action of the powder upon the
infide of the piece muft not only be fufficient to ftrain the metal,
and produce a motion in its parts, but this effect muft be extremely rapid; and the heat will be much augmented, if the
exertion of the force and the duration of its action are momen-
taneous; for in tnat cafe, the fibres of the metal (if I may ufe
the expreffion) that are violently ftretched, will return with
their full force and velocity, and the fwift vibratory motion and
attrition before defcribed will be produced.

The heat generated in a piece by firing is therefore as the
force by which the particles of the metal are ftrained and compreffed, the fuddennefs with which this force is exerted, and the
fhortnefs of the time of its action; that is to fay, as the
ftrength of the powder and the quantity of the charge, the
quicknefs of its inflammation, and the velocity with which the
generated fluid makes its efcape.

* Vide SHAW's tranflation of BOERHAAVE's Chemiftry, vol. I. p. 249.

Now

Now the effort of any given charge of powder upon the gun is very nearly the fame, whether it be fired with a bullet or without; but the velocity with which the generated elaftic fluid makes its efcape, is much greater when the powder is fired alone, than when it is made to impel one or more bullets; the heat ought therefore to be greater in the former cafe than in the latter, as I found by experiment.

But to make this matter ftill plainer, we will fuppofe any given quantity of powder to be confined in a fpace that is juft capable of containing it, and that in this fituation it is by any means fet on fire. Let us fuppofe this fpace to be the chamber of a piece of ordnance of any kind, and that a bullet, or any other folid body, is fo firmly fixed in the bore immediately upon the charge, that the whole effort of the powder fhall not be able to remove it. As the powder goes on to be inflamed, and the elaftic fluid is generated, the preffure upon the infide of the chamber will be increafed, till at length all the powder being burnt, the ftrain upon the metal will be at its greateft height, and in this fituation things will remain, the cohefion or elafticity of the particles of metal counterbalancing the preffure of the fluid.

Under thefe circumftances very little heat would be generated; for the continued effort of the elaftic fluid would approach to the nature of the preffure of a weight; and that concuffion, vibration, and friction, among the particles of the metal, which in the collifion of elaftic bodies is the caufe of the heat that is produced, would fcarcely take effect.

But inftead of being firmly fixed in its place, let the bullet now be moveable, but let it give way with great difficulty, and by flow degrees. In this cafe, the elaftic fluid will be generated as before, and will exert its whole force upon the chamber

of

of the piece; but as the bullet gives way to the preffure, and moves on in the bore, the fluid will expand itfelf and grow weaker, and the particles of the metal will gradually return to their former fituations; but the velocity with which the metal reftores itfelf being but fmall, the vibration that remains in the metal, after the elaftic fluid has made its efcape, will be very languid, as will be the heat that is generated by it.

But if, inftead of giving way with fo much difficulty, the bullet is much lighter, fo as to afford but little refiftance to the elaftic fluid in making its efcape, or if the powder is fired without any bullet at all; then, there being little or nothing to op-pofe the flame in its paffage through the bore, it will expand itfelf with an amazing velocity, and its action upon the gun will ceafe almoft in an inftant, the ftrained metal will reftore itfelf with a very rapid motion, and a fharp vibration will enfue, by which the piece will be much heated.

Of the effect of ramming the powder in the chamber of the piece.

The charge, confifting of 218 grains of powder, being put gently into the bore of the piece in a cartridge of very fine paper, without being rammed, the velocity of the bullets at a mean of the 40th, 41ft, 42d, and 47th experiments, was at the rate of 1225 feet in a fecond; but in the 68th, 69th, and 70th experiments, when the fame quantity of powder was rammed down with five or fix hard ftrokes of the ram-rod, the mean velocity was 1329 feet in a fecond. Now the total force or preffure exerted by the charge upon the bullet is as the fquare of its velocity, and $\overline{1329}^2$ is to $\overline{1225}^2$ as 1,1776 is to 1; or

nearly

nearly as 6 is to 5 ; and in that proportion was the force of the given charge of powder increased by being rammed.

In the 71st experiment the powder was also rammed, but the vent, instead of being at the bottom of the bore, was at 1,3, and the velocity of the bullet was very considerably diminished, being only at the rate of 1080 feet in a second, instead of 1276 feet in a second, which was the mean velocity with this charge, and with the vent in this situation when the powder was rammed. See the experiments N° 43, 44, 45, and 46.

When, instead of ramming the powder, or pressing it gently together in the bore, it is put into a space larger than it is capable of filling, the force of the charge is thereby very sensibly lessened, as Mr. ROBINS and others have found by repeated trials. In my 30th experiment the charge, consisting of no more than 165 grains of powder, was made to occupy 3,2 inches of the bore instead of 1,45 inches, which space it just filled when it was gently pushed into its place without being rammed ; the consequence was, the velocity of the bullet, instead of being 1100 feet in a second or upwards, was only at the rate of 914 feet in a second, and the recoil was lessened in proportion.

And from hence we may draw this practical inference, that the powder, with which a piece of ordnance or a fire-arm is charged, ought always to be pressed together in the bore ; and if it is rammed to a certain degree, the velocity of the bullet will be still farther increased. It is well known, that the recoil of a musket is greater when its charge is rammed than when it is not; and there cannot be a stronger proof that ramming increases the force of the powder.

Of the relation of the velocities of bullets to the charges of powder by which they are impelled.

It appears by all the experiments that have hitherto been made upon the initial velocities of bullets, that when the weights and dimenfions of the bullets are the fame, and they are difcharged from the fame piece by different quantities of powder, the velocities are in the fub-duplicate ratio of the weights of the charges very nearly.

The following table will fhew how accurately this law obtained in the foregoing experiments.

Charges.	Velocities		Difference.	N° of exp.
	Computed.	Actual.		
437½	1764	1764	——	3
330	1533	1594	+ 61	2
310	1486	1459	− 27	1
290	1436	1436	0	7
218	1232	1225	− 7	4
208	1216	1256	+ 40	3
165	1083	1087	+ 4	2
145	1018	1040	+ 22	2
104	860	757	− 103	2

The computed velocities, as they are fet down in this table, were determined from the ratio of the fquare root of 437½ (the weight in grains of the largeft charge of powder) to the mean velocity of the bullet with that charge and the vent at o; *viz.* 1764 feet in a fecond, and the fquare root of the other charges expreffed in grains. And the *actual* velocities are means of all experiments that were made under fimilar circumftances.

The

The fourth column fhews the difference of the computed and
actual velocities, or the number of feet in a fecond by which
the actual velocity exceeds or falls fhort of the computed: and
in the fifth column is fet down the number of experiments
with each charge, from the mean of which the actual velocity was determined.

The agreement of the computed and actual velocities will
appear more ftriking, if we take the fum and difference of thofe
velocities with all the charges except the firft: thus,

Sum of the velocities, − 1764.

Computed.	Actual.	Difference.	N° of exp.
9864	9854	− 10	23

So that it appears, that the difference, or the actual velocity, was fmaller than the computed by $\frac{1}{986}$ part only at a
mean of 23 experiments.

But as by far the greater number of the experiments were
made with the following charges, *viz.* 290, 218, 208, 165,
and 145 grains of powder, let us take the fum and difference
of the computed and actual velocities of thofe charges: thus,

Sum of the velocities

Computed.	Actual.	Difference.	N° of exp.
5985	6044	+ 59	18

Here the agreement of the theory with the experiments is fo
very remarkable, that we muft fuppofe it was in fome meafure
accidental; for the difference of the velocities in repeating the
fame experiment is in general much greater than the difference
of the computed and actual velocities in this inftance; but, I
think, we may fairly conclude, from the refult of all thefe

trials,

trials, that the velocities of like *mufket* bullets, when they are
difcharged from the fame piece by different quantities of the
fame kind of powder, are very nearly in the fub-duplicate
ratio of the weights of the charges. Whether this law will
hold good when applied to cannon balls, and bomb fhells of
large dimenfions, I dare not at prefent take upon me to decide ;
but, for feveral reafons that might be mentioned, I am rather
of opinion, that it will not; at leaft not with that degree of
accuracy which obtained in thefe experiments.

Of the effeft of plaing the vent in different parts of the charge.

There have been two opinions with refpeft to the manner in
which gun-powder takes fire. Mr. ROBINS fuppofes that the
progrefs of its inflammation is fo extremely rapid, " that all
" the powder of the charge is fired and converted into an elaftic
" fluid, before the bullet is fenfibly moved from its place ;"
while others have been of opinion, that the progrefs of the inflammation is much flower, and that the charge is feldom or
never completely inflamed before the bullet is out of the gun.

The large quantities of powder that are frequently blown
out of fire arms un-inflamed, feem to favour the opinion of
the advocates for the gradual firing; but Mr. ROBINS endea-
vours to account for that circumftance upon different principles,
and fupports his opinion by fhewing that every increafe of the
charge within the limits of practice produces a proportional
increafe of the velocity of the bullet, and that when the powder is confined by a great additional weight, by firing two or
more bullets at a time inftead of one, the velocity is not fenfibly greater than it ought to be according to his theory.

6

If

If this were a queftion merely fpeculative, it might not be worth while to fpend much time in the difcuffion of it; but as it is a matter upon the knowledge of which depends the determination of many important points refpecting artillery, and from which many ufeful improvements may be derived, too much pains cannot be taken to come at the truth. Till the manner in which powder takes fire, and the velocity with which the inflammation is propagated, are known, nothing can with certainty be determined with refpect to the beft form for the chambers of pieces of ordnance, or the moft advantageous fituation for the vent; nor can the force of powder, or the ftrength that is required in different parts of the gun, be afcertained with any degree of precifion.

As it would be eafy to determine the beft fituation for the vent from the velocity of the inflammation of powder being known, fo on the other hand I had hopes of being able to come at that velocity by determining the effect of placing the vent in different parts of the charge; for which purpofe the following experiments were made.

A table

A table of experiments, shewing the effect of placing the vent in different parts of the charge.

Weight of the charge of powder.	Space occupied by the powder.	Vent from the bottom of the bore.	Velocity of the bullet at a medium.	Recoil measured upon the ribbon at a medium.	Number of experiments.
Grains.	Inches.	Inches.	Ft. in a sec.	Inches.	
165	1,45	0	1087	14,465	2
. . .	. . .	1,32	1082	14,31	3
218	1,9	0	1225	17,93	4
. . .	. . .	1,3	1276	18,34	4
290	2,6	0	1427	22,626	3
. . .	. . .	1,3	1493	23,34	3
. . .	. . .	2,6	1460	23,286	4
. . .	. . .	0	1444	23,135	4
. . .	. . .	2,6	1413	24,5	3
310	2,7	0	——	24.69	1
. . .	. . .	1,32	——	24,95	1
. . .	. . .	2,65	——	24,9	1
330	2,9	0	1594	26,075	2
. . .	. . .	1,3	1625	26,4	2
. . .	. . .	2,6	1525	25,3	2
437½	3,9	0	1764	33,3	3
. . .	. . .	2,6	1751	32,866	3

By the foregoing experiments it appears, first, that the difference in the force of any given charge of powder which arises from the particular situation of the vent is extremely small.

With 165 grains of powder, and the vent at 0, the velocity of the bullet at a mean of two experiments (*viz.* the 20th and 21st) was 1087 feet in a second; and with the same charge, and the vent at 1,32 inches, the velocity at a mean of the 22d, 23d, and 24th experiments, was 1082 feet in a second; the

the difference, equal five feet in a fecond, is lefs than what occurred in a repetition of the fame experiment.

With 218 grains of powder, and the vent at o, the velocity at a mean in the 40th, 41ft, 42d, and 47th experiments, was at the rate of 1225 feet in a fecond; and with the fame charge, and the vent at 1,3, the velocity was 1276 feet in a fecond at a mean of four experiments, *viz.* the 43d, 44th, 45th, and 46th.

In the firft fet of experiments, with 290 grains of powder, the velocities were,

Vent at o.	Vent at 1,3.	Vent at 2;6.
1414	1476	1502
1455	1520	1450
1412	1483	1433
		1454
3)4281	3)4479	4)5839.
Means 1427	1493	1460

See the experiments from N° 48. to N° 57. inclufive.

In the fecond fet the velocities were,

Vent at o.	Vent at 2,6.
1419	1438
1460	1423
1462	1378.
1436	
4)5777	3)4239
Means 1444	1413

See the experiments from N° 109. to N° 115. inclufive.

And

And taking the means of all the velocities in both fets in each pofition of the vent it will be,

	Vent at o.	Vent at 1.3.	Vent at 2,6.
Mean velocity	1436	1493	1437

The mean recoils in thefe experiments were,

Vent at o.	Vent at 1,3.	Vent at 2,6.
22,88	23,34	23,61

In the experiments with 310 grains of powder the velocities of the bullets were not determined with fufficient accuracy to be depended on; but the recoils, which were meafured with great nicety, were as follows, *viz.*

Vent at o.	Vent at 1,3.	Vent at 2,6
24,69	24,95	24,9

With 330 grains of powder the mean velocities and recoils were,

	Vent at o.	Vent at 1,3.	Vent at 2,6.
Velocities	1594	1625	1525
Recoils	26,075	26,4	25,3

In the experiments with 437½ grains (an ounce avoirdupois) of powder the velocities and recoils were,

Vent at o.		Vent at 2,6.	
Velocity.	Recoil.	Velocity.	Recoil.
1738	33,	1707	32,5
1824	33,3	1757	33,2
1728	33,6	1789	32,9
3)5291	3)99,9	3)5253	3)98,6
Means 1764	33,3	1751	32,866

Secondly,

Secondly, From the result of all these experiments it appears,
that the effect of placing the vent in different positions with
respect to the bottom of the chamber is different, in different
charges ; thus, with 165 grains of powder the velocity of the
bullet was rather diminished by removing the vent from o, or
the bottom of the bore to 1,32 ; but with 218 grains of powder the velocity was a little increased, as was also the recoil.
With 290 grains of powder the velocity was greatest when the
powder was lighted at the vent 1,3 which was near the middle
of the charge, and rather greater when it was lighted at the
top, or immediately behind the bullet, than when it was
lighted at the bottom. And by the recoil it would seem, that
the velocities of the bullets varied nearly in the same manner
when the charge consisted of 310 grains of powder.

With 330 grains of powder, both the velocity and the recoil
were greater when the powder was lighted at the middle of
the charge, than when it was lighted at the bottom ; but they
were least of all when it was lighted near the top. And when
an ounce of powder was made use of for the charge, its force
was greatest when it was lighted at the bottom. But the difference in the force exerted by the powder which arose from the
particular position of the vent was in all cases so inconsiderable
(being, as I have before observed, less than what frequently
occurred in repeating the same experiment) that no conclusion
can be drawn from the experiments, except only this, that any
given charge of powder exerts nearly the same force, whatever
is the position of the vent.

And hence the following practical inference naturally occurs,
viz. that in the formation of fire-arms no regard need be had to
any supposed advantages that gun-smiths and others have hitherto imagined were to be derived from particular situations for

the

the vent, such as diminishing the recoil, increasing the force of the charge, &c.; but the vent may be indifferently in any part of the chamber where it will best answer upon other accounts: and there is little doubt but the same thing will hold good in great guns, and all kinds of heavy artillery.

Almost every workman who is at all curious in fire-arms has a particular fancy with regard to the best form for the bottom of the chamber, and the proper position of the vent. They in general agree, that the vent should be as low or far back as possible, in order, as they pretend, to lessen the recoil; but no two of them make it exactly in the same manner. Some make the bottom of the chamber flat, and bring the vent out even with the end of the breech-pin. Others make the vent slanting through the breech-pin, in such a manner as to enter the bore just in its axis. Others again make the bottom of the chamber conical; and there are those who make a little cylindric cavity in the breech-pin, of about two-tenths of an inch in diameter, and near half an inch in length, coinciding with the axis of the bore, and bring out the vent even with the bottom of this little cavity.

The objection to the first method is, the vent is apt to be stopped up by the foul matter that adheres to the piece after firing, and which is apt to accumulate, especially in damp weather. The same inconvenience in a still greater degree attends the other methods, with the addition of another, arising from the increased length of the vent; for the vent being longer it is not only more liable to be obstructed, but it takes a longer time for the flame to pass through it into the chamber, in consequence of which the piece is slower in going off, or, as sportsmen term it, is apt to hang fire.

2

The

The form I would recommend for the bottom of the bore is that of a hemifphere; and the vent fhould be brought out directly through the fide of the barrel, in a line perpendicular to its axis, and pointing to the center of the hemifpheric concavity of the chamber.

In this cafe the vent would be the fhorteft poffible; it would be the leaft liable to be obftructed, and the piece would be more eafily cleaned, than if the bottom of the bore was of any other form. All thefe advantages, and feveral others not lefs important, would be gained by making the bottom of the bore and vent of great guns in the fame manner.

A new method of determining the velocities of bullets.

From the equality of *action* and *re-action* it appears, that the *momentum* of a gun muft be precifely equal to the momentum of its charge; or that the weight of the gun, multiplied into the velocity of its recoil, is juft equal to the weight of the bullet and of the powder (or the elaftic fluid that is generated from it) multiplied into their refpective velocities: for every particle of matter, whether folid or fluid, that iffues out of the mouth of a piece, muft be impelled by the action of fome power, which power muft *re-act* with equal force againft the bottom of the bore.

Even the fine invifible elaftic fluid that is generated from the powder in its inflammation cannot put itfelf in motion without re-acting againft the gun at the fame time. Thus we fee pieces, when they are fired with powder alone, recoil as well as when their charges are made to impel a weight of fhot, though the recoil is not in the fame degree in both cafes.

It

It is eafy to determine the velocity of the recoil in any given cafe, by fufpending the gun in an horizontal pofition by two pendulous rods, and meafuring the arc of its afcent, by means of a ribbon according to the method already defcribed, and this will give the momentum of the gun, its weight being known, and confequently the momentum of its charge. But in order to determine the velocity of the bullet from the recoil, it will be neceffary to find out how much the weight and velocity of the elaftic fluid contributes to it.

That part of the recoil which arifes from the expanfion of this fluid is always very nearly the fame, whether the powder is fired alone, or whether the charge is made to impel one or more bullets, as I have found by a great variety of experiments.

If therefore a gun, fufpended according to the method prefcribed, is fired with any given charge of powder, but without any bullet or wad, and the recoil is obferved, and if the fame piece is afterwards fired with the fame quantity of powder, and a bullet of a known weight, the excefs of the velocity of the recoil in the latter cafe, over that in the former, will be proportional to the velocity of the bullet; for the difference of thefe velocities, multiplied into the weight of the gun, will be equal to the weight of the bullet multiplied into its velocity.

Thus if W is put equal to the weight of the gun,

U = the velocity of its recoil, when it is fired with any given charge of powder, without any bullet,

V = the velocity of the recoil, when the fame charge is made to impel a bullet,

B = the weight of the bullet, and

v = its velocity,

It will be $v = \dfrac{\overline{V-U} \times W}{B}$.

I

Let us fee how this method of determining the velocities of bullets will anfwer in practice.

In the 94th experiment the recoil, with 165 grains of powder, without a bullet, was 5,5 inches, and in the 95th experiment, with the fame charge, the recoil was 5,6 inches. The mean is 5,55 inches; and the length of the rods by which the barrel was fufpended being 64 inches, the velocity of the recoil (= U) anfwering to 5,55 inches meafured upon the ribbon, is that of 1,1358 feet in a fecond.

In five experiments, with the fame charge of powder, and a bullet weighing 580 grains, the recoil was as follows, *viz.*

The 20th experiment   14,73 inches
    21ft   -   14,2
    22d   -   14,8
    23d   -   14,58
    24th   -   14,68

$$5)73, \quad (= 14,6 \text{ inches at a mean.}$$

And the velocity of the recoil (= V) anfwering to the length is that of 2,9880 feet in a fecond: confequently V – U, or 2,9880 – 1,1358 is equal to 1,8522 feet in a fecond.

But as the velocities of recoil are known to be as the chords of the arcs through which the barrel afcends, it is not neceffary in order to determine the velocity of the bullet to compute the velocities V and U; but the quantity $\overline{V - U}$, or the difference of the velocities of the recoil when the given charge is fired with and without a bullet, may be computed from the value of the difference of the chords, by one operation. Thus the velocity anfwering to the chord 9,05 is that of 1,8522 feet in a fecond, which is juft equal to $\overline{V - U}$, as was before found.

The

The weight of the barrel, together with its carriage, was 47¼ pounds, to which three quarters of a pound is to be added on account of the weight of the rods by which it was fuspended, which makes W = 48 pounds, or 336,000 grains, and the weight of the bullet was 580 grains. B is therefore to W as 580 is to 336,000, that is, as 1 is to 579,31 very nearly; and v ($= \dfrac{\overline{V-U} \times W}{B}$ is equal to $\overline{V-U} \times 579,31$.

The value of $\overline{V-U}$ answering to the experiments before mentioned was found to be 1,8522; consequently the velocity of the bullets ($= v$) was $1,8522 \times 579,31 = 1073$ feet in a second, which is extremely near 1083 feet in a second, the mean of the velocities, as they were determined by the pendulum.

But the computation for determining the velocity of a bullet upon these principles may be rendered still more simple and easy in practice; for the velocities of the recoil being as the chords measured upon the ribbon, if

> c is put equal to the chord of the recoil expressed in English inches, when the piece is fired with powder only, and
>
> C = the chord when a bullet is discharged by the same charge,

then C – c will be as V – U; and consequently as $\dfrac{\overline{V-U} \times W}{B}$, which measures the velocity of the bullet, the ratio of W to B remaining the same.

If therefore we suppose a case in which C – c is equal to one inch, and the velocity of the bullet is computed from that chord, the velocity in any other case, wherein C – c is greater or less than one inch, will be found by multiplying the difference of the chords C and c by the velocity that answers to a difference of one inch.

The

The length of the parallel rods by which the barrel was fufpended being 64 inches, the velocity of the recoilanfwering to
$C - c = 1$ inch meafured upon the ribbon is 0,204655 parts of a
fo   in a fecond; and this is alfo, in this cafe the value of
$V - U$; the velocity of the bullet is therefore $v = 0,204655 \times$
$579,31 = 118,35$ feet in a fecond.

Confequently the velocity of the bullet expreffed in feet *per*
fecond may in all cafes be found by multiplying the difference
of the chords C, and c, by 118,35, the weight of the barrel, the
length of the rods by which it is fufpended, and the weight of
the bullet remaining the fame, and this whatever the charge of
powder may be that is made ufe of, and however it may differ
in ftrength or goodnefs.

According to this rule the velocities of the bullets in the following experiments have been computed from the recoil; and
by comparing them with the velocities fhewn by the pendulum,
we fhall be enabled to judge of the accuracy of this new method of determining the velocities of bullets.

In the 76th and 77th experiments, when the piece was fired
with 145 grains of powder and a bullet, the recoil was 13,25
and 13,15, or 13,2 at a mean; and with the fame charge of
powder, without a bullet, the recoil was 4,5 and 4,3, or 4,4
at a mean (fee the 84th and 93d experiments). $C - c$ is
therefore $13,2 - 4,4 = 8,8$ inches, and the velocity of the bullets $= 8,8 \times 118,35 = 1045$ feet in a fecond. The mean of
the velocities as they were determined by the pendulum is that
of 1040 feet in a fecond. In the 104th and 105th experiments
the recoil was 12,92 and 13,28, and the velocity computed
from the mean of thofe chords is 1030 feet in a fecond; but
the velocity fhewn by the pendulum was no more than about
900 feet in a fecond. As the recoil was fo nearly equal to
what

what it was in the 76th and 77th experiments before mentiond, when the velocities fhewn by the recoil and by the pendulum were almoft exactly the fame, I am inclined to believe, that there muft have been fome miftake in determining the velocities by the pendulum in thefe laft experiments, and that the velocity fhewn by the recoil is moft to be depended on.

With 290 grains, or half the weight of the bullet in powder, in the 48th, 49th, and 50th experiments, the recoil was 22,58, 22,92, and 22,38; and the recoil, with the fame charge of powder, without a bullet, at a mean of the 60th and 99th experiments, was 10,66. The mean of the velocities of the bullets, computed from the recoil, is therefore 1416 feet in a fecond, and the velocity fhewn by the pendulum was 1427 feet in a fecond: the difference is not confiderable. The mean of the velocities in the 109th, 110th, 111th, and 112th experiments is by the recoil 1464, and by the pendulum 1444 feet in a fecond.

With 330 grains of powder the velocities of the bullets appear to have been as follows, *viz.*

	Vent at o.	Vent at 1,3.	Vent at 2,6.
By the recoil	1543	1620	1610
By the pendulum	1594	1625	1528

See the 62d, 63d, 64th, 65th, 66th, 67th, and 17th experiments.

The uniformity of the recoil was in all cafes very remarkable. Thus, in the firft fet of experiments with 290 grains of powder (from the 48th to the 57th experiment inclufive), the recoil was,

Vent at o.	Vent at 1,3.	Vent at 2,6.
22,58	23,21	23,06
22,92	23,76	23,26
22,38	23,06	23,26
		23,56
3)67,88	3)70,03	4)93,14

Means = 22,626   23,343   and   23,285

If now we take a mean of the 60th and 99th experiments, and call the recoil, without a bullet, 10,66 as before, the velocities will turn out,

	Vent at o.	Vent at 1,3.	Vent at 2,6.
By the recoil - -	1416	1501	1494
And by the pendulum they were	1427	1493	1494
The difference is only .	− 11	+ 8	+ 34

The recoil was equally regular in the 117th and five succeeding experiments, when the charge was no lefs than 437½ grs. = 1 ounce avoirdupois in powder; and the velocities of the bullets determined from the recoil are very nearly the fame as they were fhewn by the pendulum. Thus, in the 117th, 118th, and 119th experiments the mean recoil was 33,3; and in the 120th, 121ft and 122d experiments it was 32,866. And if the recoil without a bullet is called 17,9, as it was determined by the 123d experiment, which was made immediately after the experiments before mentioned, then will the velocities be,

	Vent at o.	Vent at 2,6
By the recoil - -	1822	1771
And by the pendulum they were	1764	1751

The difference is only   - $+58$ and $+20$ feet in a fecond, which is lefs than what frequently occurs in repeating the fame experiment.

In the 11th, 12th, 13th, and 14th experiments, when the piece was fired with 310 grains of powder and a bullet, the recoil was 24,69, 24,95, 24,9, and 24,9 : and in the 15th, 16th, 18th, and 19th experimentss with 330 grains of powder, the recoil was 26,2, 26,2, 26,3, and 26,4. The regularity of thefe numbers is very ftriking; and though we cannot compare the velocities of the bullets determined by the two methods as we have done in other cafes (as there are reafons to believe, that the velocities, as they are fet down in the tables, are not much to be depended on, and as the recoil, with the given charge of 310 grains of' powder without a bullet, is not known) yet the regularity of the recoil in thefe experiments affords good grounds to conclude, that the method of determining the velocities of bullets founded upon it muft be very accurate.

But of all the experiments thofe numbered from 84 to 92 inclufive afford the ftrongeft proof of the accuracy of this method. In thofe every poffible precaution was taken to prevent errors arifing from adventitious circumftances, and the weights of the bullets and their velocities were fo various, that the uniform agreement of the two methods of determining the velocities in thofe trials amounts almoft to a demonftration of the truth of the principles upon which this new method is founded.

By

By the following table the refult of thefe experiments may be feen at one view.

The experiments.	Weight of the bullets.	The barrel heavier than the bullet.	The recoil.	Velocity of the bullet.		Difference.
				By the recoil.	By the pendulum	
	Grs.	$\dfrac{W}{B} =$	$C =$	$v =$	$v =$	
84th and 93d	——	—— $c =$	4,4	——	——	——
85th	90	3733,3	7,16	2109	1763	+ 346
86th	251	1338,6	9,62	1430	1317	+ 113
87th	354	949,15	11,03	1288	1136	+ 152
88th	600	560,	15,22	1240	1229	+ 11
89th	603	557,22	15,13	1224	1229	+ 5
90th	1184	283,78	21,92	1017	978	− 39
91ft	1754	191,56	27,18	893	916	− 23
92d	2352	141,86	32,25	812	833	− 21

The charge of powder confifted of 145 grains in weight in each experiment.

In order to fhew, in a more ftriking manner, the refult of thefe experiments, and the comparifon of the two methods of afcertaining the velocities of bullets, I have drawn the fig. 16. where the numbers that are marked upon the line AB are taken from A towards B, in proportion to the weights of the bullets; while the lines drawn from thofe numbers perpendicular from AB (as *w*, *v*, for inftance, at the number 2352) and ending at the curve *c*, *d*, exprefs their velocities, as fhewn by the pendulum. The continuation of thofe lines on the oppofite fide of the line AB fhew the recoil, and alfo the velocities of the bullets as determined from it; thus *w*, *r*, and the (dotted) lines

2 Q 2   parallel

parallel to it, which end at the line *g*, *f*, exprefs the recoil ;
and the portion of each of thofe lines that is comprehended between the line AB and the curve *m*, *n* (as *w*, *u*) is as the velocity of the bullet in the feveral experiments. The line A, *c*,
denotes the weight of the charge of powder ; and the line
A, *m*, the velocity with which the elaftic fluid efcapes out of
the piece, when the powder is fired without any bullet.

Upon an infpection of this figure, as well as from an examination of the foregoing table, it appears, that the velocities
determined by the two methods agree with great nicety in all
the experiments after the 87th ; but in the 87th experiment,
and alfo in the 86th, but particularly in the 85th, the difference in the refult of thefe different methods is very confiderable : and it is remarkable, that in thofe experiments where
they difagree moft, the velocities of the bullets, as determined
by the pendulum, are extremely irregular ; while, on the other
hand, the gradual increafe of the recoil as the bullets were
heavier, and the great regularity of the correfponding velocities, afford good grounds to conclude, that this difagreement is
not owing to any inaccuracy in the new method of afcertaining
the velocities, but to fome other caufe that remains to be in-
veftigated.

But before we proceed in this inquiry, let us feparate the five
laft experiments in the foregoing table ; and, fumming up the
velocities determined by the two methods, we fhall fee by their
difference how thofe methods agreed upon the whole, in this
inftance.

Experiments.	Weight of the bullets.	Velocity.		Difference.
		By the recoil.	By the pendulum.	
	Grs.			
88th	600	1240	1229	+ 11
89th	603	1224	1229	− 5
90th	1184	1017	978	+ 39
91ſt	1754	893	916	− 23
92d	2352	812	833	− 21
Sums and diff. of the velocities		5186	5185	+ 1

Here the difference in the reſult of the two methods does not amount to $\frac{1}{1000}$th part of the whole velocity; but I lay no ſtreſs upon this extraordinary argument. I am ſenſible that it muſt in ſome degree have been accidental ; but as the difference in the velocities, computed by theſe different methods, was in no inſtance conſiderable, not being in any caſe ſo great as what frequently occurred in the moſt careful repetition of the ſame experiment, and as the velocities, as determined by the recoil, were much more regular than thoſe ſhewn by the pendulum, as appears by compaing the curves *g*, *f*, and *m*, *n*, (fig. 16.) with the crooked line *c*, *d*, I think we may fairly conclude, that this new method may with ſafety be relied on in practice.

The greateſt difference in the velocities, as aſcertained by the two methods, appears, in the inſtance of the 85th experiment, where the velocity determined from the recoil exceeds that ſhewn by the pendulum by 346 feet in a ſecond, the former velocity being that of 2109 feet in a ſecond, the latter only 1763 feet in a ſecond ; and in the two ſucceeding experiments, the velocities ſhewn by the pendulum are likewiſe deficient, though not in ſo great a degree.

This

This apparent deficiency remains now to be accounted for ; and, firſt, it cannot be ſuppoſed, that it aroſe from any imperfection in Mr. ROBINS's method of determining the velocities of bullets ; for that method is founded upon ſuch principles as leave no room to doubt of its accuracy ; and the practical errors that occur in making the experiments, and which cannot be intirely prevented, or exactly compenſated, are in general ſo ſmall, that the difference of the velocities in queſtion cannot be attributed to them. It is true, the effect of thoſe errors is more likely to appear in experiments made under ſuch circumſtances as thoſe under which the experiments we are now ſpeaking of were made, than in any other caſe ; for the bullets being very light, the arc of the aſcent of the pendulum was but ſmall, and a ſmall miſtake in meaſuring the chord upon the ribbon would have produced a very conſiderable error in computing the velocity of the bullet : thus, a difference of one tenth of an inch, more or leſs, upon the ribbon in the 85th experiment, would have made a difference in the velocity of more than 120 feet in a ſecond. But independent of the pains that were taken to prevent miſtakes, the ſtriking agreement of the velocities determined by the two methods in the experiments which immediately follow, as alſo in all other caſes where they could be compared, affords abundant reaſon to conclude, that the errors ariſing from thoſe cauſes were in no inſtance very conſiderable.

But if both methods of aſcertaining the velocities of bullets are to be relied on, then the difference of the velocities, as determined by them in theſe experiments, can only be accounted for by ſuppoſing that it aroſe from their having been diminiſhed by the reſiſtance of the air in the paſſage of the bullets from the mouth of the piece to the pendulum ; and this ſuſpicion

will

will be much ftrengthened when we confider how great the
refiftance is that the air oppofes to bodies that move very fwiftly
in it, and that the bullets in thefe experiments were not only
projeﬅed with great velocities, but were alfo very light, and
confequently more liable to be retarded by the refiftance on that
account.

To put the matter beyond all doubt, let us fee what the refiftance was that thefe bullets met with, and how much their
velocities were diminifhed by it.   The weight of the bullet (in
the 85th experiment) was 90 grains; its diameter was 0,78 of
an inch, and it was projeﬅed with a velocity of 2109 feet in a
fecond.

If now a computation be made according to the method laid
down by Sir ISAAC NEWTON for compreffed fluids, it will be
found, that the refiftance to this bullet was not lefs than $8\frac{1}{4}$
lbs. avoirdupois, which is fomething more than 660 times its
weight.   But Mr. ROBINS has fhewn, by experiment, that
the refiftance of the air to bodies moving in it with very great
velocity is near three times greater than Sir ISAAC has
determined it, and as the velocity with which this bullet was
impelled is confiderably greater than any in Mr. ROBINS's experiments, it is highly probable, that the refiftance in this inftance was at leaft 2000 times greater than the weight of the
bullet.

The diftance from the mouth of the piece to the pendulum,
as we have before obferved, was 12 feet; but, as there is reafon to think, that the blaft of the powder, which always follows the bullet, continues to aﬅ upon it for fome fenfible fpace
of time after it is out of the bore, and by urging it on coun-
ter-balances, or at leaft counter-aﬅs in a great meafure the
refiftance of the air, we will fuppofe, that the refiftance does

I   not

not begin, or rather that the motion of the bullet does not begin to be retarded, till it has got to the diftance of two feet from the muzzle. The diftance, therefore, between the barrel and the pendulum, inftead of 12 feet, is to be eftimated at 10 feet; and as the bullet took up about $\frac{1}{145}$ part of a fecond in running over that fpace, it muft in that time have loft a velocity of about 335 feet in a fecond, as will appear upon making the computation, and this will very exactly account for the apparent diminution of the velocity in the experiment; for the difference of the velocities, as determined by the recoil and by the pendulum, $= 2109 - 1763 = 346$ feet in a fecond, is extremely near 335 feet in a fecond, the diminution of the velocity by the refiftance as here determined.

If the diminution of the velocities of the bullets in the two fubfequent experiments be computed in like manner, it will turn out in the 86th experiment $= 65$ feet in a fecond, and in the 87th experiment $= 33$ feet in a fecond; and making thefe corrections, the comparifon of the two methods of afcertaining the velocities will ftand thus:

	85th exp.	86th exp.	87th exp.
Velocities fhewn by the pendulum,	1763	1317	1136
Add the diminution of the velocity by the refiftance of the air,	335	65	33
	2098	1382	1169
Velocity by the recoil,	2109	1430	1288
The difference =	+11	+48	+119

So that it appears, notwithftanding thefe corrections, that the velocities in the 86th and 87th experiments, and particularly in the laft, as they were determined by the pendulum, are

ftill

ftill confiderably deficient. But the manifeft irregularity of the velocities in thofe inftances affords abundant reafon to conclude, that it muft have arifen from fome extraordinary accidental caufe, and therefore, that little dependance is to be put upon the refult of thofe experiments. I cannot take upon me to determine pofitively what the caufe was which produced this irregularity; but I ftrongly fufpect, that it arofe from the breaking of the bullets in the barrel by the force of the explofion: for thefe bullets, as has already been mentioned, were formed of lead, inclofing leffer bullets of plafter of Paris; and I well remember to have obferved at the time feveral fmall fragments of the plafter which had fallen down by the fide of the pendulum. I confefs, I did not then pay much attention to this circumftance, as I naturally concluded, that it arofe from the breaking of the bullet in penetrating the target of the pendulum, and that the fmall pieces of plafter I faw upon the ground had fallen out of the hole by which the bullet entered. But if the bullets were not abfolutely broken in pieces in firing, yet, if they were confiderably bruifed, and the plafter or a part of it were feparated from the lead, fuch a change in their form might produce a great increafe of the refiftance, and even their initial velocities might be affected by it, for their form being changed from that of a globe to fome other figure, they might not fit the bore, and a part of the force of the charge might be loft by the windage.

That this actually happened in the 87th experiment feems very probable, as the velocity with which the bullet was projected, as it was determined by the recoil, is confiderably lefs in proportion in that experiment than in either of thofe that precede it in that fet, or in thofe which follow it, as will ap-

pear upon infpecting the curvature of the line *m*, *n*, fig. 16:.
But I forbear to infift further upon this matter.

As I have made an allowrance for the refiftance of the air in
thefe experiments, it may be expected that I fhould do it in all
other cafes; but, I think, it will appear upon enquiry, that
the diminution of the velocities of the bullets on that account
was in general fo inconfiderable that it might fafely be neg-
lected : thus, for inftance, in the experiments with an ounce
of powder, when the velocity of the bullet was more than
1750 feet in a fecond, the diminution turns out no more than
about 25 or 30 feet in a fecond, though we fuppofe the full
refiftance to have begun fo near as two feet from the mouth of
the piece ; and in all cafes where the velocities were lefs, the
effect of the refiftance was lefs in a much greater proportion :
and even in this inftance there is reafon to think, that the diminution of the velocity as we have determined it is too great ;
for the flame of gun-powder expands with fuch an amazing
rapidity, that it is fcarcely to be fuppofed but that it follows
the bullet, and continues to act upon it more than two feet, or
even four feet, from the gun, and when the velocity of the
bullet is lefs, its action upon it muft be fenfible at a ftill
greater diftance.

With 218 grains of powder the recoil appears to have been
very uniform ; but if the velocities of the bullets are determined from the recoil in the 40th and feven following experiments, when this charge was made ufe of, and from the recoil
without a bullet in the 72d and 73d experiments, the velocities
will turn out confiderably too fmall, as we fhall fee by making
the computation.

The

	Vent at 0.				Vent at 1,3.
	⌠ 40th exp. was 18,	and in the	43d exp.	it was 18,3
The re-	41ft -	17,71	-	44th	- 18,35
coil in the	42d -	17,91	-	45th	- 18,35
	⌊ 47th -	8,1	-	46th	- 18,35

$$4)71,72 \qquad\qquad 4)73,35$$

Means = 17,93   and   18,34

And in the 72d and 73d experiments the recoil, with the fame charge without a bullet, was 8,72, and 8,47 = 8,595 at a medium, the velocities therefore turn out,

Vent at 0.   Vent at 1,3.

By the recoil 1105 - 1153
inftead of   1225 and 1276 as they were fhewn by the pend.

The difference 120 and 123 feet in a fecond amounts to near one twelfth part of the whole velocity.

This difference is undoubtedly owing to the recoil without a bullet being taken too great; for it is not only greater than it ought to be, in order that the velocities of the bullets may come out right; but it is confiderably greater in proportion than the recoil with any other charge.

Thus, with 145 grains of powder the recoil was 4,4
        with 165 grains   -   -   it was 5,55
             290 grains   -   -   -   10,66
             330 grains   -   -   -   12,7
   and with 437½ grains   -   -   - it was 17,9

And if the recoil with 218 grains is determined from thefe numbers by interpolation, it comes out 7,5; and with that

value for C the velocities of the bullets in the before mentioned experiments appear to be,

	Vent at o.		Vent at 1,3.	
	1243	and	1283	by the recoil
which is extremely near	1225	and	1276	the velocities

fhewn by the pendulum.

It is to be remembered, that the 72d and 73d experiments, from
which we before determined the recoil with the given charge
of powder without a bullet, were not made upon the fame day
with the experiments before mentioned ; and it is well known,
that the force of powder is different upon different days. And
it is worthy of remark, that in thofe two experiments the
ftrength of government powder appeared to be confiderably the
greateft. I mention thefe circumftances to fhew the probability
there is, that the recoil in thofe experiments, from fome un-
known caufe, was greater than it ought to have been, or rather than it would have been, had the experiments been made
at the fame time when the experiments with the bullets were
made ; or at any other time under the fame circumftances.

As this method of determining the velocities of the bullets
did not occur to me till after I had finifhed the courfe of my
experiments, and had taken down my apparatus, I have not
had an opportunity of afcertaining the recoil, with and without a bullet, with that degree of precifion that I could wifh.
If I had thought of it fooner, or if I had recollected that paffage in Mr. ROBINS's new Principles of Gunnery, where he
fays, " The part of the recoil, arifing from the expanfion of
" the powder alone, is found to be no greater when it impels
" a leaden bullet before it, than when the fame quantity is fired
" without any wad to confine it :" I fay, if that paffage had
occurred

occurred to me before it had been too late, I certainly fhould have taken fome pains to have afcertained the fact; but as it is, I think, enough has been done to fhew, that there is the greateft probability that the velocities of bullets may in all cafes be determined by the recoil with great accuracy; and I hope foon to have it in my power to put the matter out of all doubt, and to verify this new method by a courfe of conclufive experiments which I am preparing for that purpofe.

In the mean time I would juft obferve, that if this method fhould be found to anfwer, when applied to mufket bullets, it cannot fail to anfwer equally well when it is applied to cannon balls and bomb fhells of the largeft dimenfions; and it is apprehended, that it will be much preferable to any method hitherto made public; not only as it may be applied indifferently to all kinds of military projectiles, and that with very little trouble or expence in making the experiment; but alfo becaufe by it the velocities with which bullets are *actually projected* are determined; whereas by the pendulum their velocities can only be afcertained at fome diftance from the gun, and after they have loft a part of their initial velocities by the refiftance of the air through which they are obliged to pafs to arrive at the pendulum.

'At the trifling expence of ten or fifteen pounds an apparatus might be conftructed that would anfwer for making the experiments with all the different kinds of ordnance in the Britifh fervice. The advantages that might be derived from fuch a fet of experiments are too obvious to require being mentioned.

Of a very accurate method of proving gun-powder.

All the *eprouvettes*, or powder-triers, in common ufe are defective in many refpects. Neither the abfolute force of gun-
powder can be determined by means of them, nor the comparative force of different kinds of it, but under circumftances
very different from thofe in which the powder is made ufe of
in fervice.

As the force of powder arifes from the action of an elaftic
fluid that is generated from it in its inflammation, the quicker
the charge takes fire, the more of this fluid will be generated
in any given fhort fpace of time, and the greater of courfe will
be its effect upon the bullet. But in the common method of
proving gun-powder, the weight by which the powder is confined is fo great in proportion to the quantity of the charge,
that there is time quite fufficient for the charge to be all inflamed, even when the powder is of the floweft compofition,
before the body to be put in motion can be fenfibly removed
from its place. The experiment, therefore, may fhew which
of two kinds of powder is the ftrongeft, when equal quantities
of both are confined in equal fpaces, and completely inflamed;
but the degree of the inflammability, which is a property effen-
tial to the goodnefs of the powder, cannot by thefe means be
afcertained.

Hence it appears, how powder may anfwer to the proof,
fuch as is commonly required, and may neverthelefs turn out
very indifferent when it comes to be ufed in fervice. And this,
I believe, frequently happens; at leaft I know complaints
from officers of the badnefs of our powder are very common;
and I would fuppofe that no powder is ever received by the

6   Board

Board of Ordnance but fuch as has gone through the eſta-
bliſhed examination, and has anſwered to the uſual teſt of its
being of the ſtandard degree of ſtrength.

But though the common powder triers may ſhew powder
to be better than it really is, they never can make it appear to
be worſe than it is; it will therefore always be the intereſt of
thoſe who manufacture that commodity to adhere to the old
method of proving it. But the purchaſer will find his account
in having it examined in a method by which its goodneſs may
be aſcertained with greater preciſion.

The method I would recommend is as follows. A quantity
of powder being provided, which, from any previous examination or trial, is known to be of a proper degree of ſtrength to
ſerve as a ſtandard for the proof of other powder, a given charge
of it is to be fired, with a fit bullet, in a barrel ſuſpended by
two pendulous rods, according to the method before deſcribed,
and the recoil is to be carefully meaſured upon the ribbon. And
this experiment being repeated three or four times; or oftener if
there is any difference in the recoil, the mean and the extremes
of the chords may be marked upon the ribbon by black lines
drawn acroſs it, and the word *proof* may be written upon the
middle line; or if the recoil is uniform (which it will be to a
ſufficient degree of accuracy, if care is taken to make the experiments under the ſame circumſtances) then the *proof mark*
is to be made in that part of the ribbon to which it was con-
ſtantly drawn out by the recoil in the different trials.

The recoil, with a known charge of ſtandard powder, being
thus aſcertained and marked upon the ribbon, let an equal
quantity of any other powder (that is to be proved) be fired in
the ſame barrel, with a bullet of the ſame weight, and every
other circumſtance alike, and if the ribbon is drawn out as
far

far or farther than the proof mark, the powder is as good or better than the standard; but if it falls short of that distance, it is worse than the standard, and is to be rejected.

For the greater the velocity is with which the bullet is impelled, the greater will be the recoil; and when the recoil is the same, the velocities of the bullets are equal, and the powder is of the same degree of strength, if the quantity of the charge is the same. And if care is taken in proportioning the charge to the weight of the bullet, to come as near as possible to the medium proportion that obtains in practice, the determination of the goodness of gun-powder from the result of this experiment cannot fail to hold good in actual service.

Fig. 14. represents the proposed apparatus, drawn to a scale of one foot to the inch. *a, b,* is the barrel suspended by the pendulous rods *c, d*; and *r* is the ribbon for measuring the recoil.

The length of the bore is 30 inches, and its diameter is one inch, consequently it is just 30 calibres in length, and will carry a leaden bullet of about 3 ounces.

The barrel may be made of gun metal, or of cast iron as that is a cheaper commodity; but great care must be taken in boring it, to make the cylinder perfectly strait and smooth, as well as to preserve the proper dimensions. Of whatever metal the barrel is made, it ought to weigh at least 50 lbs. in order that the velocity of the recoil may not be too great; and the rods by which it is suspended should be five feet in length. The vent may be about one twentieth of an inch in diameter; and it should be *bouched* or lined with gold, in the same manner as the touch-hole is made in the better kind of fowling pieces, in order that its dimensions may not be increased by repeated firing.

The

The bullets fhould be made to fit the bore with very little windage; and it would be better if they were all caft in the fame mould, and of the fame parcel of lead, as in that cafe their weights and dimenfions would be more accurately the fame, and the experiments would of courfe be more conclufive.

The ftated charge of powder may be half an ounce, and it fhould always be put up in a cartridge of very fine paper; and after the piece is loaded it fhould be primed with other powder, firft taking care to prick the cartridge by thrufting a priming wire down the vent.

As it appears, from feveral experiments made on purpofe to afcertain the fact, that ramming the powder more or lefs has a very fenfible effect to increafe or diminifh the force of the charge; to prevent any inaccuracies that might arife from that caufe, a ram-rod, fuch as is reprefented fig. 15. may be made ufe of. It is to be made of a cylindric piece of wood in the fame manner as ram-rods in general are made, but with the addition of a ring C, about one inch and a half, or two inches in diameter, which, being placed at a proper diftance from the end (*a*) of the ram-rod that goes up into the bore, will prevent its being thruft up too far. This ring may be made of wood, or of any kind of metal as fhall be found moft convenient. The other end of the ram-rod (*b*) may be 31 or 32 inches in length from the ring, and the extremity of it being covered with a proper fubftance, it may be made ufe of for wiping out the barrel after each experiment.

The machine (*f*) for the tape to flide through may be the fame as is defcribed by Dr. HUTTON in his account of his experiments on the initial velocities of cannon balls; as his method is much better calculated to anfwer the purpofe than that propofed and made ufe of by Mr. ROBINS. It will alfo be

better for the axis of the pendulous rods to reſt upon level
pieces of wood or iron, than for them to move in circular
grooves : only care muſt be taken to confine them by ſtaples or
ſome other contrivance, to prevent their ſlipping out of their
places.

The trunnions, by means of which the barrel is connected
with the pendulous rods, and upon which it is ſupported,
ſhould be as ſmall as poſſible, in order to leſſen the friction ;
and for the ſame reaſon they ſhould be well poliſhed, as well as
the grooves that receive them.   They need not be caſt upon the
barrel, but may be ſcrewed into it after it is finiſhed.

In making the experiments, regard muſt be had to the heat
of the barrel, as well as to the temperature of the atmoſphere ;
for heat and cold, dryneſs and moiſture, have a very ſenſible
effect upon gun-powder to increaſe or diminiſh its force.   If
therefore a very great degree of accuracy is at any time required, it will be beſt to begin by firing the piece two or three
times merely to warm it ; after which three or four experiments may be made with ſtandard powder, to determine anew
the proof mark (for the ſtrength of the ſame powder is different upon different days) ; and when this is done, the experiments with the powder that is to be proved are to be made,
taking care to preſerve the ſame interval of time between the
firings, that the heat of the piece may be the ſame in each
trial.

If all theſe precautions are taken, and if the bullets are of
the ſame weight and dimenſions, powder may be proved by
this method with much greater accuracy than has hitherto been
done by any of the common methods made uſe of for that
purpoſe.

Of the comparative goodnefs, or value, of powder of different degrees of ftrength.

Let V denote the velocity of the bullet with the ftronger powder, and put *v* equal to the velocity with the weaker, when the charges are equal, and the weight and dimenfions of the bullets are the fame, and when they are difcharged from the fame piece. If the charge is augmented when the weaker powder is made ufe of, till the velocity of the bullet is increafed from *v* to V, or becomes equal to the velocity with the given charge of the ftronger powder, the *value* of the charges may then be faid to be equal; and confequently the weaker powder is as much worfe than the ftronger, or is of lefs value in proportion as the quantity of it required by the pound, to produce the given effect is greater.

But it is well known, that the velocities, with different quantities of the fame kind of powder, are in the fub-duplicate ratio of the weights of the charges. The charges, therefore, muft be as the fquares of the velocities, and confequently the charge of the weaker powder muft be to that of the ftronger, when the velocities are equal, as VV is to *vv*. The weaker powder is therefore as much worfe than the ftronger as VV is greater than *vv*; or the comparative goodnefs of powder, of different degrees of ftrength, is as the fquares of the velocities of the bullets when the charges are equal.

The mean velocity of the bullets, as fhewn by the pendulum in the 104th and 105th experiments, when the piece was fired with 145 grains of government powder, was 894 feet

S f 2   in in a fecond; and with the fame quantity of double proof * battle powder (experiment N° 106) the velocity was 990 feet in a fecond. Now the fquares of thefe velocities, which, as we juft obferved, meafure the goodnefs of the powder, are to each other as 1 is to 1,2263, or nearly as 5 is to 6.

With 218 grains of government powder, the mean velocity in four experiments *(viz.* the 40th, 41ft, 42d, and 43d) was 1225 feet in a fecond; and in the experiment N° 107. when the fame quantity of double proof battle powder was made ufe of, the velocity was 1380 in a fecond; and $\overline{1225}^2$ is to $\overline{1380}^2$ as 1 is to 1,2691.

With 290 grains, or half the weight of the bullet in government powder in the 109th, 110th, 111th, and 11⸱th experiments, the mean velocity of the bullet was 1444 feet in a fecond; but with the fame quantity of the battle powder (experiment N° 116.) the velocity was 1525 feet in a fecond; $\overline{1444}^2$ is to $\overline{1525}^2$, as 1 is to 1,1153.

By taking a medium of thefe trials it appears, that double proof battle powder is better than government powder in proportion as 1,2036 is to 1, or nearly as 6 is to 5.

But if, inftead of weighing the powder, we eftimate the quantity of the charge by meafurement, or the fpace it occupies in the bore of the piece, the comparative ftrength of battle powder will appear to be confiderably greater, or its ftrength will be to that of government powder nearly as 4 is to 3 ; for the grains of this better kind of powder being more compact and nearly of a fpherical form, a greater weight of it will lie in any

---

* This is called *battle* powder, not becaufe it is ufed in battle or in war; but from *Battle*, the name of a village in Kent, where that kind of powder is made.

given fpace than of government powder, which is formed more loofely, and of various and of very irregular figures.

Now the common price of double proof battle powder, as it is fold by the wholefale dealers in that commodity, is at the rate of £. 10 *per* cwt. net, which is juft two fhillings by the pound ; while government powder is fold at £. 5 5*s. per* hundred, or one fhilling and $\frac{6}{10}$th of a penny *per* pound ; but battle powder is better than government powder only in the proportion of 1,2036 to 1, or of one fhilling and two pence to one fhilling and $\frac{6}{10}$th of a penny ; battle powder is therefore fold at the rate of ten pence by the pound, or 41$\frac{2}{1}$. *per cent.* dearer than it ought to be ; or thofe, who make ufe of it in preference to government powder, do it at a certain lofs of 41$\frac{2}{1}$ *per cent.* of the money that the powder cofts them.

Of the relation of the velocities of bullets to their weights.

According to Mr. ROBINS's theory, when bullets of the fame diameter, but different weights, are difcharged from the fame piece by the fame quantity of powder, their velocities fhould be in the reciprocal fub-duplicate ratio of their weights ; but as this theory is founded upon a fuppofition that the action of the elaftic fluid, generated from the powder, is always the fame in any and every given part of the bore when the charge is the fame, whatever may be the weight of the bullet ; and as no allowance is made for the expenditure of force required to put the fluid itfelf in motion, or for the lofs of it by the vent ; it is plain that the theory is defective. It is true, Dr. HUTTON in his experiments found this law to obtain without any great error, and poffibly it may hold good with fufficient accuracy in many cafes ; for it fometimes happens that a number of errors or
                                                            actions,

actions, whofe operations have a contrary tendency, fo compen-
fate each other, that their effects when united are not fenfible.
But when this is the cafe, if any one of the caufes of error is
removed, thofe which remain will be detected.

When any given charge is loaded with a heavy bullet, more
of the powder is inflamed in any very fhort fpace of time than
when the bullet is lighter, and the action of the powder ought
of courfe to be greater on that account; but then a heavy bullet takes up more time in paffing through the bore than a light
one, and confequently more of the elaftic fluid, generated from
the powder, efcapes by the vent and by windage. It may happen, that the augmentation of the force, on account of one of
thefe circumftances, may exactly counterbalance the diminution
of it arifing from the other; and if it fhould be found upon
trial that this is the cafe in general, in pieces as they are now
conftructed, and with all the variety of fhot that are made ufe
of in practice, it would be of great ufe to know the fact: and
poffibly it might anfwer as well, as far as it relates to the art of
gunnery, as if we were perfectly acquainted with, and were
able to appreciate, the effect of each varying circumftance under
which an experiment can be made. But when, concluding too
haftily from the refult of a partial experiment, we fuppofe
with Mr. ROBINS, that becaufe the fum total of the action or
preffure of the elaftic fluid upon the bullet, during the time of
its paffage through the bore, happens to be the fame when bullets of different weights are made ufe of (which collective
preffure is in all cafes proportional to, and is accurately meafured by, the velocity, or rather motion, communicated to the
bullet) that therefore the preffure in any given part is always
exactly the fame, when the quantity of powder is the fame
with which the piece is fired; and from thence endeavour to

prove,

prove, that the inflammation of gun-powder is inftantaneous, or that the whole charge is in all cafes inflamed, and " con- " verted into an elaftic fluid before the bullet is fenfibly moved " from its place ;" fuch reafonings and conclufions may lead to very dangerous errors.

It is undoubtedly true, that if the principles affumed by Mr. ROBINS with refpect to the manner in which gun-powder takes fire, and the relation of the clafticity of the generated fluid to its denfity, or the intenfity of its preffure upon the bullet as it expands in the barrel, were juft, and if the lofs of force by the vent and windage was in all cafes inconfiderable, or if it was prevented, the deductions from the theory refpecting the velo- cities of bullets of different weights would always hold good. But if, on the contrary, it fhould be found upon making the experiments carefully, and in fuch a manner as intirely to pre- vent inaccuracies arifing from adventitious circumftances, that the velocities obferve a law different from that which the theory fuppofes, we may fairly conclude, that the principles upon which the theory is founded are erroneous.

Let us now fee how far thefe experiments differ from the theory. Thofe numbered from 84 to 92 inclufive were made in fuch a manner that no part of the force of the powder was loft by the vent, or by windage, as has already been mentioned, and all poffible attention was paid to every circumftance that could contribute to render them perfect and conclufive.

A particular account of them with the means ufed for form- ing the bullets, and making them fit for the bore, and the con- trivance for preventing the efcape of the elaftic fluid by the vent, &c. may be feen in the general table, p. 245. The fol- lowing table fhews the refult of them.

3   N. B.

N. B. The charge of powder was the fame in each experiment, and confifted of 145 grains in weight.

	Weight of the bullet.	Velocity of the bullet.		Difference.
		Actual.	Computed.	
85th exp.	90 grs.	2109	2109	————
86th	251	1430	1262	+ 168
87th	354	1288	1063	+ 225
88th	600	1240	817	+ 423
89th	603	1224	815	+ 409
90th	1184	1017	581	+ 436
91ft	1754	893	478	+ 415
92d	2352	812	413	+ 399

The computed velocities, as they are fet down in this table, were determined from the actual velocity of the bullet, as determined by the recoil in the 85th experiment; and the reciprocal fub-duplicate ratio of its weight to the weight of the bullet in each fubfequent experiment; and in the laft column is marked the difference between the experiment and the theory, or the number of feet in a fecond, by which the actual velocity exceeds the computed.

But in order that we may fee this matter in different points of view, let the order of the experiments be now inverted, and let the computed velocities be determined from the actual velocity in the 92d experiment; and affuming the total or collective preffure exerted by the powder upon the bullet in that experiment equal to unity, let the collective preffure in the other experiments be computed from the ratio of the actual to the computed velocities, and the table will ftand thus:

	Weight of the bullet.	Velocity of the bullet		Difference.	Collective pressure.
		Actual.	Computed.		
92d exp.	2352 grs.	812	812	———	1,0000
91ft	1754	893	940	— 47	0,9020
90th	1184	1017	1145	— 128	0,7897
89th	603	1224	1604	— 330	0,5825
88th	600	1240	1608	— 368	0,5949
87th	354	1288	2093	— 805	0,3778
86th	251	1430	2486	— 1056	0,3310
85th	90	2109	4151	— 2042	0,2581

In the following figure let AB reprefent the axis of the piece, and AP the length of the fpace filled with powder; and at the

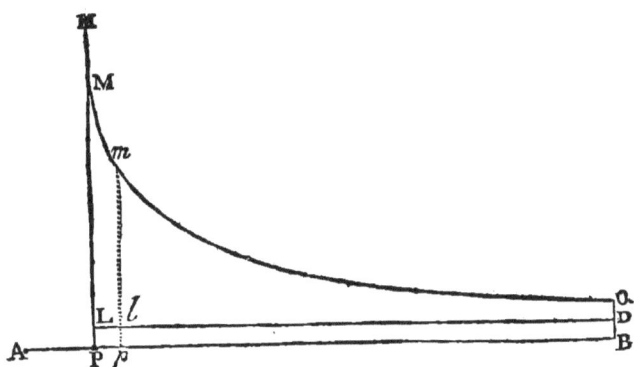

point P let the perpendicular PH be erected, upon which let PL and PM be taken from P towards H of fuch magnitudes that while PL expounds the uniform force of gravity, or the weight of the bullet, PM fhall be as the force exerted by the powder

upon the bullet at the moment of the explosion. If now we
suppose, that while the bullet moves on from P towards B, the
line PM or *pm*, goes along with it, and that the point *m* is
always taken in such a manner that the line *pm*, shall be to
pl, or PL, as the force acting upon the bullet in the point *p*,
is to its weight, till *pm*, coincides with QB, then will the
area PMQB be to the area PLDB in the duplicate proportion of
the velocities which the bullet would acquire when acted on by
its own gravity through the space PB, and when impelled
through the same space by the force of the powder, as may
be seen demonstrated by Sir ISAAC NEWTON, in his Mathematical Principles of Natural Philosophy, book I. prop. 39.

Now what I call the collective pressure, or sum total of the
action of the powder upon the bullet, is the measure of the
area PMQB; and it is plain, from what has been said above,
that its measures are in all cases to be accurately determined,
when the weight and velocity of the bullet are known.

If all the powder of the charge was inflamed at once, or before the bullet sensibly moved from its place ; and if the pressure of the generated fluid was always as its density, or inversely
as the space it occupies, then would the line MQ be an hyberbola,
the area PMQB would always be the same when the charge
was the same, and consequently the velocities of the bullets
would be as the square roots of their weights inversely. But
it appears, from the before mentioned experiments, that when
the weight of the bullet was increased four times, the action
of the powder, or area PMQB, was nearly doubled ; for in
the 92d experiment, when four bullets were discharged at
once, the collective pressure was as 1 ; but in the 89th experiment, when a single bullet was made use of, the collective

I   pressure

preffure was only as 0,5825; and in the 85th, 86th, and 87th experiments, when the bullets were much lighter, the action of the charge was ftill lefs.

But though we can determine with great certainty, from thefe experiments, the ratio in which the action of the powder upon the bullet was increafed or diminifhed, by making ufe of bullets of greater or lefs weight; yet we cannot from thence afcertain the relation of the elafticity of the generated fluid to its denfity, nor the quantity of powder that is inflamed at dif-ferent periods before and after the bullet begins to move in the bore.

But affuming Mr. ROBINS's principles as far as relates to the elafticity of the fluid, and fuppofing that in all the experiments, except the 92d, a part only of the charge took fire, and that that part was inflamed and converted into an elaftic fluid before the bullet began to move; upon that fuppofition we can deter-mine the quantity of powder that took fire in each experiment; for the quantity of powder in that cafe would be as the collec-tive preffure.

Thus, if the whole charge, = 145 grains in weight, is fup-pofed to have been inflamed in the 92d experiment, the quan-tity inflamed in each of the other experiments will appear to have been as follows; *viz.*

T t 2　　　　　　　　　Weight

	Weight of the bullet.	Velocity of the bullet.	Collective preffure.	Powder inflamed.
85th exp.	90 grs.	2109	0,2581	37 grs.
86th	251	1430	0,3310	48
87th	354	1288	0,3778	55
88th	600	1240	0,5949	86
89th	603	1224	0,5825	84
90th	1184	1017	0,7897	114
91ft	1754	893	0,9020	131

But there are many reafons to fuppofe, that the diminution
of the action of the powder upon the bullet, when it is lighter, is
not fo much owing to the fmallefs of the quantity of powder
that takes fire in that cafe as to the *vis inertiæ* of the generated
fluid. It is true, that a greater portion of the charge takes
fire when the bullet is heavy than when it is light, as I found
in the very experiments of which I am now fpeaking; but then
the quantity of unfired powder in any cafe was much too fmall
to account for the apparent diminution of the force when light
bullets were made ufe of.

If the elaftic fluid, in the action of which the force of
powder confifts, were infinitely fine, or if its weight bore no
proportion to that of the powder that generated it; and if the
grofs matter, or *caput mortuum,* of the powder remained in the
bottom of the bore after the explofion, then, and upon no
other fuppofition, would the preffure upon the bullet be inverfely as the fpace occupied by the fluid: but it is evident that
this can never be the cafe.

A curious fubject for fpeculation here occurs: how far would
it be advantageous, were it poffible, to diminifh the fpecific
gravity of gun-powder, and the fluid generated from it, with-

3,   out

out leſſening its elaſtic force? It would certainly act upon very light bullets with greater force; but when heavy ones came to be made uſe of, there is reaſon to think, that, except extraordinary precaution was taken to prevent it, the greateſt part of the force would be loſt by the vent and by windage.

The velocity with which elaſtic fluids ruſh into a void ſpace is as the elaſticity of the fluid directly, and inverſely as its denſity; if, therefore, the denſity of the fluid generated from powder was four times leſs than it is, its elaſticity remaining the ſame, it would iſſue out at the vent, and eſcape by the ſide of the bullet in the bore, with nearly four times as great a velocity as it does at preſent; but we know from experiment that the loſs of force on thoſe accounts is now very conſiderable.

In the experiments N° 76. and 77. when the piece was fired with 145 grains of powder, the velocity of the bullets at a medium was 1040 feet in a ſecond; but in the 88th and 89th experiments, when the bullets were even heavier, and the piece was fired with the ſame quantity of powder, the mean velocity was 1232 feet in a ſecond. The difference = 192 feet in a ſecond, anſwers to a difference of force greater in the laſt experiments than in the firſt in the proportion of 14 to 10.

I know of no way to account for this difference, but by ſuppoſing that it was owing intirely to the eſcape of the elaſtic fluid by the vent, and by windage, in thoſe experiments where the vent was open, and the bullets were put naked into the piece.

An elaſtic bow, made of very light wood, will throw an arrow, and eſpecially a light one, with greater velocity than a bow of ſteel of the ſame degree of ſtiffneſs: but, for practice, I think it is plain, that gun-powder may *be ſuppoſed* to be ſo

light

light as to be rendered intirely ufelefs : and for fome purpofes it
feems probable, that it would not be the worfe for being even
heavier than it is now made.   Vents are abfolutely neceffary in
fire-arms, and in large pieces the windage muft be confiderable,
in order that the bullets, which are not always fo round as they
fhould be, may not ftick in the bore ; and thofe who have been
prefent at the firing of heavy artillery and large mortars with
fhot and fhells, muft have obferved, that there is a fenfible fpace
of time elapfes between the lighting of the prime and the explofion ; and that, during that interval, the flame is continually
iffuing out at the vent with a hiffing noife, and with a prodigious velocity, as appears by the height to which the ftream of
fire mounts up in the air.   It is plain, that this lofs muft be
greater in proportion as the fhot that is difcharged is heavier ;
and I have often fancied, that I perceived a fenfible difference in
the time that elapfed between the firing of the prime and the
explofion, when bullets were difcharged, and when the piece
has been fired with powder only ; the time being apparently
longer in the former cafe than the latter.

Almoft all the writers upon gun-powder, and particularly
thofe of the laft century, gave different recipes for powder that
is defigned for different ufes.   Thus the French authors mention *poudre à moufquet*, *poudre ordinaire de guerre*, *poudre de
chaffe*, and *poudre d'artifice* ; all of which are compofed of falt
petre, fulphur, and charcoal, taken in different proportions.
Is it not probable, that this variety in the compofition of powder was originally introduced in confequence of obfervation
that one kind of powder was better adapted for particular
purpofes than another, or from experiments made on purpofe
to afcertain the fact ?   There is one circumftance that would
lead us to fuppofe that that was the cafe.— That kind of pow-
                                                                der

der which was defigned for great guns and mortars was weaker
than thofe which were intended to be ufed in fmaller pieces :
for if there is any foundation for thefe conjectures, it is certain, that the weakeft powder, or the heavieft in proportion to
its elaftic force, ought to be ufed to impel the heavieft bullets,
and particularly in guns that are imperfectly formed, where the
vent is large, and the windage very great.

I am perfectly aware, that an objection may here be made,
viz. that the elaftic fluid, which is generated from gun-powder,
muft be fuppofed to have the fame properties very nearly,
whatever may be the proportion of the feveral ingredients, and
that therefore the only difference there can be in powder is,
that one kind may generate more of this fluid, and another lefs ;
and that when it is generated, it acts in the fame manner, and
will alike efcape, and with the fame velocity, by any paffage it
can find. But to this I anfwer, though the fluid may be the
fame, as undoubtedly it is, and though its denfity and elafticity may be the fame in all cafes at the inftant of its generation,
yet in the explofion, the elaftic and un-elaftic parts are fo mixed
and blended, that I imagine the fluid cannot expand without
taking the grofs matter along with it, and the velocity with
which the flame iffues out at the vent is to be computed from
the elafticity of the fluid, and the denfity or weight of the fluid
and the grofs matter taken together, and not fimply from
the elafticity and denfity of the fluid. If antimony in an im-
palpable powder, or any other heavy body, was intimately
mixed with water in a veffel of any kind, and kept in fufpenfion
by fhaking or ftirring them about ; and if a hole was opened
in the fide or bottom of the veffel, the water would not run
out without taking the particles of the folid body along with
it. And in the fame manner I conceive the folid particles that
remain

remain after the explofion of gun-powder are carried forward with the generated elaftic fluid, and being carried forward retard its motion.— But to return from this digreffion.

As it appears from thefe experiments, that the relation of the velocities of bullets to their weights is different from that which Mr. ROBINS's theory fuppofes, it remains to inquire what the law is which actually obtains. And, firft, as the velocities bear a greater proportion to each other than the reciprocal fub-duplicate ratio of the weights of the bullets, let us fee how near they come to the reciprocal fub-triplicate ratio of their weights.

	Weight of the bullet.	Velocity of the bullet				
		Computed.		Actual.		Computed.
		Recip. fub-dup. ratio.	Error of the theory.		Error of the theory.	Recip. fub-trip. ratio.
92d	2352	812	———	812	———	812
91ft	1754	940	+ 47	893	+ 2	895
90th	1184	1145	+ 128	1017	+ 4	1021
89th	603	1604	+ 380	1224	+ 54	1278
88th	600	1608	+ 368	1240	+ 40	1280
87th	354	2093	+ 805	+288	+239	1527
86th	251	2486	+1056	1430	+282	1712
85th	90	4151	+2042	2109	+301	2410

Here the velocities computed upon the laft fuppofition appear to agree much better with the experiments than thofe computed upon Mr. ROBINS's principles; but ftill there is a confiderable difference between the actual and the computed velocities in the three laft experiments in the table.

As the powder itfelf is heavy, it may be confidered as a weight that is put in motion along with the bullet; and if we

fuppofe

suppose the density of the generated fluid is always uniform from the bullet to the breech, the velocity of the center of gravity of the powder or (which amounts to the same thing) of the elastic fluid, and the gross matter generated from it will be just half as great as the velocity of the bullet. If therefore we put P to denote the weight of the powder, B the weight of the bullet, and v its initial velocity; then $Bv + \frac{1}{2}Pv = \overline{B + \frac{1}{2}P} \times v$ will express the *momentum* of the charge at the instant when the bullet quits the bore.

If now, instead of ascertaining the relation of the velocities to the weights of the bullets, we add half the weight of the powder to the weight of the bullet, and compute the velocities from the reciprocal sub-triplicate ratio of the quantity $\overline{B + \frac{1}{2}P}$ in each experiment, the table will stand thus:

	Weight of the bullet and half the powder.	Velocity of the bullet.		Error of the theory.
		Actual.	Computed.	
	$\overline{B + \frac{1}{2}P} =$			
92d exp.	$2352 + 72\frac{1}{2}$	812	812	———
91ft	$1754 + 72\frac{1}{2}$	893	892	− 1
90th	$1184 + 72\frac{1}{2}$	1017	1011	− 6
89th	$603 + 72\frac{1}{2}$	1224	1243	+ 19
88th	$600 + 72\frac{1}{2}$	1240	1245	+ 5
87th	$354 + 72\frac{1}{2}$	1288	1449	+161
86th	$251 + 72\frac{1}{2}$	1430	1589	+159
85th	$90 + 72\frac{1}{2}$	2109	1999	−110

The agreement between the actual and computed velocities is here very remarkable, and particularly in the five first experiments, which are certainly those upon which the greatest dependence may be placed.

And hence we are enabled to determine the natures of the *mn,* and *gf* (fig. 16.).; for since B (which expresses the weight of the bullet) is as the length taken from A towards B in the figure in the several experiments; and as the velocities are as the lines drawn perpendicular to the line AB from the places where those lengths terminate, as *w, u,* &c. ending at the curve *m, n;* if we put $a = \frac{1}{4}$P, $x = $B, and $y = wu$; then will the relation of x and y be defined by this equation. $\frac{1}{\sqrt{a+x^3}} = y$. And if z be put to denote the line *w r,* and *o,* the recoil when the given charge is fired without any bullet, it will be $\frac{x}{\sqrt{a+x^3}} + b = z$ in the curve *gf,* x being the abscissa, and z the corresponding ordinate to the curve.

In the 92d experiment half the weight of the powder ($= a$) was $72\frac{1}{2}$ grains; the weight of the bullet was 2352 grains ($= x$); the recoil ($= z$) was 32,25 inches, and with the given charge without any bullet the recoil ($= b$) was 4,4 inches; if now from these *data,* and the known weight of the bullet in each of the other experiments in this set, the recoil be computed by means of the theorem $\frac{x}{\sqrt{a+x^3}} + b = z$, we shall see how the result of those experiments agrees with this theory: thus,

	Weight of the bullet.	Recoil		Difference.
		Actual.	Computed.	
92d exp.	2352	32,25	32,25	——
91ft	1754	27,18	27,22	+0,04
90th	1184	21,92	21,85	−0,07
89th	603	15,13	15,33	+0,2
88th	600	15,22	15,29	+0,07
87th	354	11,03	11,87	+0,84
86th	251	9,62	10,21	+0,59
85th	90	7,16	7,02	−0,14
84th and 93d	0	4,4	4,4	——

Here the agreement of the actual and computed recoils is as remarkable as that of the actual and computed velocities in the foregoing table.

By the figure 17. may be seen at one view the result of all these experiments and computations. The numbers upon the line AB (as in the fig. 16.) represent the weights of the bullets, while the lines drawn from those numbers perpendicular to AB on each side, and ending at the curves *m*, *n*, are as the velocities of the bullets in the several experiments; the line AB being the axis of the curves, the lengths taken from A to the different numbers towards B ($= x$) the abscissa, and the perpendiculars ($= y$) the corresponding ordinates. The ordinates to the the curve *bn*, are as the velocities computed from the theorem $\frac{1}{\sqrt{a+x^3}} = y$, and the ordinates to the curve *p*, *n* (which is the logarithmic curve, as it is $\frac{1}{\sqrt{x}} = y$) shew the velocities computed upon Mr. ROBINS's principles. The curve *gf* is drawn from

U u 2   the

the theorem $-\dfrac{x}{\sqrt{a+x^3}}+b=z$; and the actual recoil is marked
upon the ordinates to this curve by large round dots, which in
all the experiments, except the 86th and 87th, very nearly
coincide with the curve.

In the fig 18. the numbers upon the line AB, taken from A,
denote the different charges of powder used in the course of
the experiments, while the ordinates to the curve cd, expres-
the velocities of the bullets, with the vent at o. The lines
drawn perpendicular from the line AB to the line ef, represent the recoil with the several charges of powder, and a leaden
bullet; and the portion of those lines that is comprehended
between the line AB and the line gb, denotes the recoil when
the given charge was fired without any bullet.

Having now shewn by experiment the relation of the velocities of bullets to their weights, when care is taken to prevent
intirely the loss of force by the escape of the elastic fluid
through the vent and by the windage, I shall leave it to ma-
thematicians to determine from these *data* the properties of
that fluid.

But, before I take my leave of this subject, I would just ob-
serve, that Mr. ROBINS is not only mistaken in the principle he
assumes, respecting the relation of the elasticity of the fluid
generated from gun-powder to its density, or rather the law of
its action upon the bullet as it expands in the bore; but his
determination of the force of gun-powder is also erroneous,
even upon his own principles: for he determines its force to be
1000 times greater than the mean pressure of the atmo-
sphere; whereas it appears, from the result of the 92d experiment, that its force is at least 1308 times greater than the
mean

mean preffure of the atmofphere, as will be evident to thofe who will take the trouble to make the computation.

Of an attempt to determine the explofive force of aurum fulmi-nans, *or a comparifon between its force and that of gun-powder.*

Having provided myfelf with a fmall quantity of this won-derful powder, upon the goodnefs of which I could depend, I endeavoured to afcertain its explofive force by making ufe of it inftead of gun-powder for difcharging a bullet, and meafuring, by means of the pendulum, the velocity which the bullet ac-quired; and concluding, from the tremendous report with which this fubftance explodes, that its elaftic force was vaftly greater than that of gun-powder, I took care to have a barrel provided of uncommon ftrength, on purpofe for the experiment. Its length in the bore is 13,25 inches, the diameter of the bore is 0,55 of an inch, and its weight 7 lbs. 2 oz. It is of the beft iron, and was made by WOGDON; and the accuracy with which it is finifhed does credit to the workman.

This barrel being charged with one fixteenth of an ounce (= 27,34 grains) of *aurum fulminans* and two leaden bullets, which, together with the leather that was put about them to make them fit the bore without windage, weighed 427 grains; it was laid upon a chaffing-difh of live coals, at the diftance of about 10 feet from the pendulum, and againft the center of the target of the pendulum the piece was directed.

Having fecured the barrel in fuch a manner that its direction fhould not alter, I retired to a little diftance, in order to be out of danger in cafe of an accident, where I waited in anxious expectation the event of the explofion.

I had

I had remained in this fituation fome minutes, and almoft defpaired of the experiment's fucceeding, when the powder exploded, but with a report infinitely lefs than what I expected, the noife not greatly exceeding the report of a well-charged wind gun; and it was not till I faw the pendulum in motion that I could be perfuaded that the bullets had been difcharged. I found, however, upon examination, that nothing was left in the barrel, and from the great number of fmall particles of revived metal that were difperfed about, I had reafon to think that all the powder had exploded.

The bullets ftruck the pendulum nearly in the center of the target, and both of them remained in the wood: and I found, upon making the calculation, that they had impinged againft it with a velocity of 428 feet in a fecond.

If we now fuppofe that the force of *aurum fulminans* arifes from the action of an elaftic fluid that is generated from it in the moment of its explofion, and that the elafticity of this fluid, or rather the force it exerts upon the bullet as it goes on to expand, is always as its denfity, or inverfely as the fpace it occupies; then, from the known dimenfions of the barrel, the length of the fpace occupied by the charge (which in this experiment was 0,47 of an inch), and the weight and velocity of the bullets, the elaftic force of this fluid at the inftant of its generation may be determined: and I find, upon making the calculation upon thefe principles, that its force turns out 307 times greater than the mean elaftic force of common air.

According to Mr. ROBINS's theory, the elaftic force of the fluid generated from gun-powder in its inflammation is 1,000 times greater than the mean preffure of the atmofphere; the force of *aurum fulminans*, therefore, appears to be to that of gun-powder as 307 is to 1,000, or as 4 is to 13 very nearly.

Of

Of the fpecific gravity of gun-powder.

To determine the fpecific gravity of gun-powder I made ufe of the following method. A large glafs bucket, with a narrow mouth, being fufpended to one of the arms of a very nice balance, and exactly counter-poifed by weights put in the oppofite fcale, it was filled firft with government powder poured in lightly, then with the fame powder fhaken well together, afterwards with powder and water together, and laftly with water alone, and in each cafe the contents of the bucket were very exactly weighed.

The fpecific gravity of gun-powder, as determined from thefe experiments, is as follows :

Specific gravity of rain water - -	1,000
Government powder, as it lies light in a heap, mixed with air - - - -	0,836
Government powder well fhaken together -	0,937
The folid fubftance of the powder - -	1,745

Hence it appears, that a cubic inch of government powder fhaken well together weighs juft 243 grains; that a cubic inch of folid powder would weigh 442 grains; and, confequently, that the interftices between the particles of the powder, as it is grained for ufe, are nearly as great as the fpaces which thofe particles occupy.

MISCELLANEOUS EXPERIMENTS.

Of some unsuccessful attempts to increase the force of gun-powder.

It has been suppofed by many, that the force of fteam is even greater than that of gun-powder; and that if a quantity of water, confined in the chamber of a gun, could at once be rarified into fteam, it would impel a bullet with prodigious ve-locity. Several attempts have been made to fhoot bullets in this manner; but I know of none that have fucceeded; at leaft fo far as to render it probable that water can ever be fubftituted in the room of gun-powder for military purpofes, as fome have imagined.

The great difficulty that attends making thefe experiments lies in finding out a method by which the water can at once be rarified, and converted into elaftic fteam; and it occurred to me, that poffibly that might be effected by means of gun-pow-der, by confining à fmall quantity of water in fome very thin fubftance, and furrounding and inclofing it with powder, and afterwards fetting fire to the charge. The method I took to do this was as follows. Having procured a number of air blad-ders of very fmall fifhes, I put different quantities of water into them from the fize of a fmall pea to that of a fmall piftol bullet, and tying them up clofe with fome very fine thread, I hung up thefe little globules in the open air till they were quite dry on the outfide. I then provided a number of cartridges made of fine paper, and filled them with a known quantity of powder, equal to the cuftomary charge for a common horfe-man's piftol; and having loaded fuch a piftol with one of them

and

and a fit bullet, I laid it down upon the ground, and directing it against an oaken plank that was placed about fix feet from the muzzle, I fired it off by a train, and carefully obferved the recoil, and alfo the penetration of the bullet. I then took feveral of the filled cartridges that remained, and pouring out part of the powder, I put one or more of the little bladders filled with water in the center of the cartridge, and afterwards pouring back the remaining part of the charge, confined the water in the midft of the powder.

With thefe cartridges and a fit bullet, the piftol was fucceffively loaded, and being placed upon the ground as before, and fired by a train, the recoil, and the penetration of the bullets were obferved; and I conftantly found, that the force of the charge was very fenfibly diminifhed by the addition of the globule of water, and the larger the quantity of water was that was thus confined, the lefs was the effect of the charge; neither the recoil of the piftol, nor the penetration of the bullet, being near equal to what they were when the given quantity of powder was fired without the water; and the report of the explofion appeared to be leffened in a ftill greater proportion than the recoil or penetration.

Concluding that this diminution of the force of the charge arofe from the burfting of the little bladder, and the difperfion of the water among the powder before it was all inflamed, by which a great part of it was prevented from taking fire, I repeated the experiments with highly rectified fpirits of wine inftead of water; but the refult was nearly the fame as before: the force of the charge was conftantly and very fenfibly diminifhed. I afterwards made ufe of etherial oil of turpentine, and then of fmall quantities of quickfilver; but ftill with no better fuccefs. Every thing I mixed with the powder, inftead of increafing, ferved to leffen the force of the charge.

Thefe trials were all made feveral months before I began the courfe of my experiments upon gun-powder, which I have already given an account of; and though they were altogether unfuccefsful, yet I refumed the inquiry at that time, and made feveral new experiments, with a view to find out fomething that fhould be ftronger than powder, or which, when mixed with it, fhould increafe its force.

It is well known, that the elaftic force of quickfilver con- verted into vapour is very great; this fubftance I made ufe of in my former trials, as I have juft obferved, but without fuc- cefs. I thought, however, that the failure of that attempt might poffibly be owing to the quickfilver being too much in a body, by which means the fire could not act upon it to the greateft advantage; but that if it could be divided into exceed- ing fmall particles, and fo ordered that each particle might be completely furrounded by, and expofed to, the action of the flame of the powder, it would be very foon heated, and poffibly might be converted into an elaftic fteam or vapour, before the bullet could be fenfibly removed from its place. To determine this point I mixed 20 grains of æthiops mineral very intimately with 145 grains of powder, and charging the piece with this compound, it was loaded with a fit bullet and fired; but the force of the charge was lefs than that which the powder alone would have exerted, as appears by comparing the 76th and 77th experiments with the 79th.

Common *pulvis fulminans* is made of one part of fulphur, two parts of falt of tartar, and three parts of nitre; and if we may judge by the report of the explofion, the elaftic force of this compound is confiderably greater than that of gun-powder.

I was

I was willing to fee the effect of mixing falt of tartar with gun-powder, and accordingly having provided fome of this alkaline falt in its pureft ftate, thoroughly dry, and in a fine powder, I mixed 20 grains of it with 145 grains of gun-powder; and upon difcharging a bullet with the mixture, I found that the alkaline falt had confiderably leffened the force of the powder. See experiment N° 78.

I next made ufe of *fal ammoniacum.* That falt has been found to produce a very large quantity of elaftic air, or vapour, when expofed to heat under certain circumftances; but when 20 grains of it were mixed with a charge of gun-powder, inftead of adding to its force, it diminifhed it very fenfibly. See the 80th experiment.

Moft, if not all, the metals, are thought to produce large quantities of air when they are diffolved in proper *menftrua,* and particularly brafs, when it is diffolved in fpirit of nitre. De-firous of feeing if this could be done by the flame, or acid vapour of fired powder, I mixed 20 grains of brafs in a very fine powder, commonly called brafs duft (being the fmall particles of this metal that fly off from the wheel in fharpening pins), with 145 grains of powder, and with this compound and a fit bullet I loaded my barrel and difcharged it; but the experiment (N° 81.) fhewed, that the force of the powder was not increafed by the addition of the brafs duft, but the con-trary.

It feems probable, however, that neither brafs duft nor æthiops mineral are of themfelves capable of diminifhing the force of gun-powder in any confiderable degree, etherwife than, by filling up the interftices between the grains, and obftructing the paffage of the flame, and fo impeding the progrefs of the in-flammation. And hence it appears, how earthy particles and

impurities

impurities of all kinds are fo very detrimental to gun-powder. It is not that they deftroy or alter the properties of any of the bodies of which the powder is compofed, but fimply, that by obftructing the progrefs of the inflammation, they leffen its force, and render it of little or no value. Too much care, therefore, cannot be taken in manufacturing powder to free the materials from all heterogeneous matter.

Of an attempt to fhoot flame inflead of bullets.

Having often obferved paper and other light bodies to come out of great guns and fmall arms inflamed, I was led to try if other inflammable bodies might not be fet on fire in like manner, and particularly inflammable fluids; and I thought if this could be effected, it might be poffible to project fuch ignited bodies by the force of the explofion, and by that means communicate the fire to other bodies at fome confiderable diftance; but in this attempt I failed totally. I never could fet dry tow on fire at the diftance of five yards from the muzzle of my barrel. I repeatedly difcharged large wads of tow and paper, thoroughly foaked in the moft inflammable fluids, fuch as *alkohol, ætherial fpirit of turpentine, balfam of fulphur,* &c.; but none of them were ever fet on fire by the explofion. Sometimes I difcharged three or four fpoonfuls of the inflammable fluid, by interpofing a very thin wad of cork over the powder, and another over the fluid; but ftill with no better fuccefs. The fluid was projected againft the wall as before, and left a mark where it hit; but it never could be made to take fire; fo I gave up the attempt. If it had fucceeded, probably it would have turned out one of the moft important difcoveries in the art of war that have been made fince the invention of gun-powder.

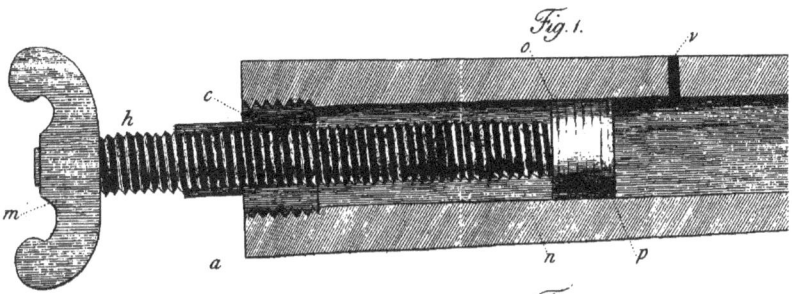

Fig. 1.

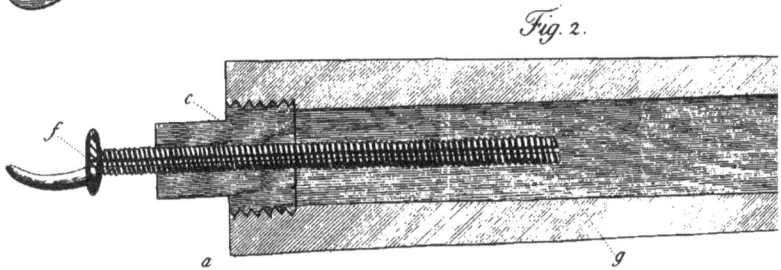

Fig. 2.

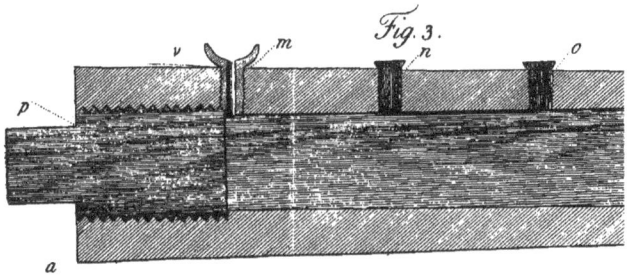

Fig. 3.

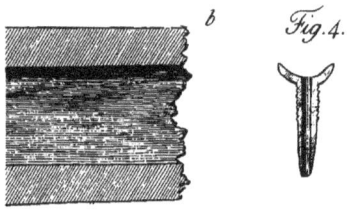

Fig. 4.

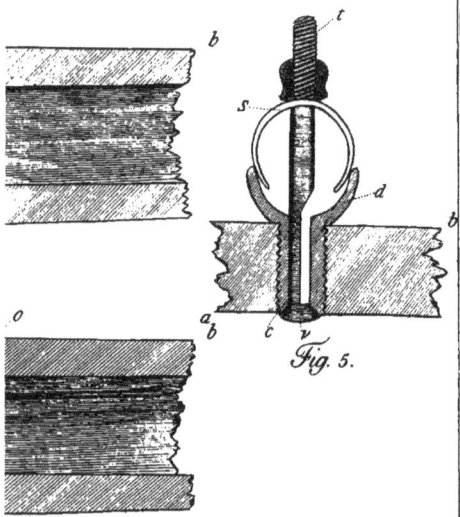

Fig. 5.

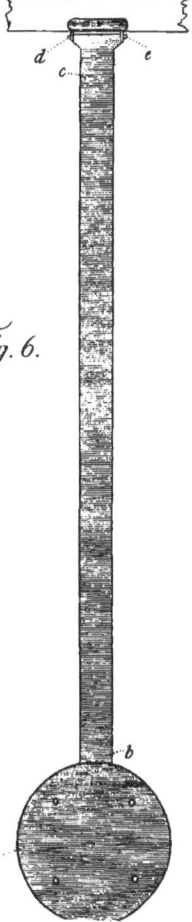

Fig. 6.

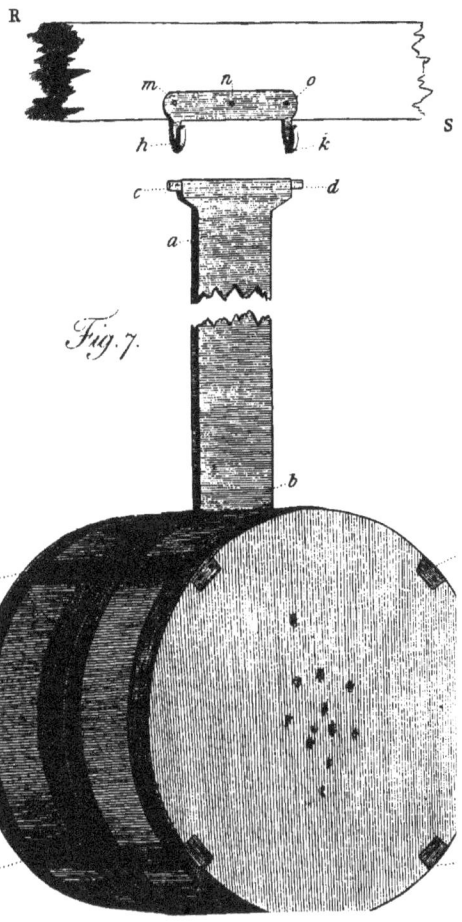

Fig. 7.

Fig. 8.

4

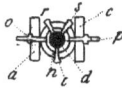

Fig. 9.

Fig. 10.

Fig. 12.

Fig.11.   *Fig.13.*   *Fig.15.*

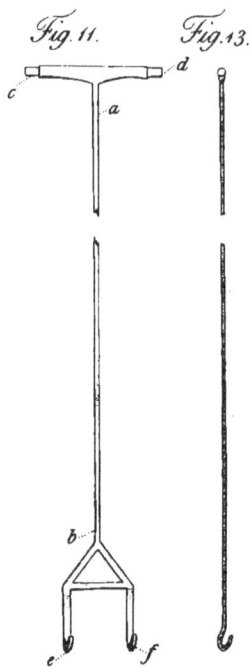
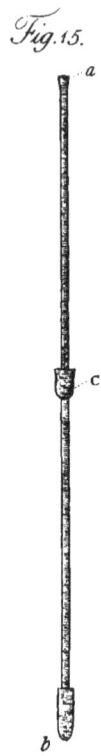

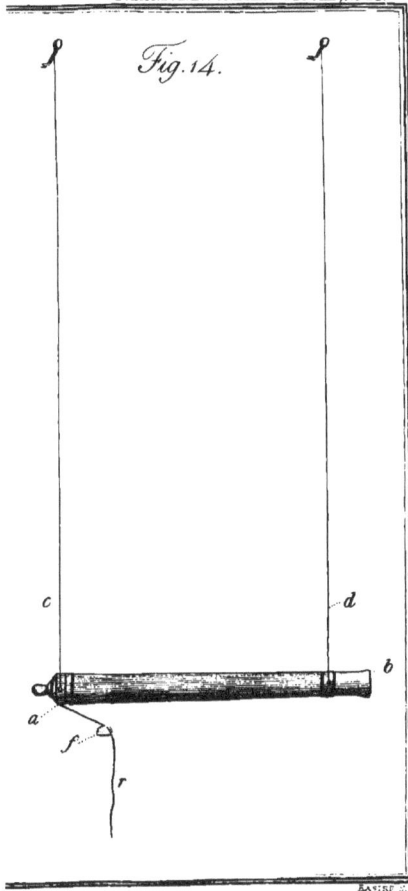

Fig.14.

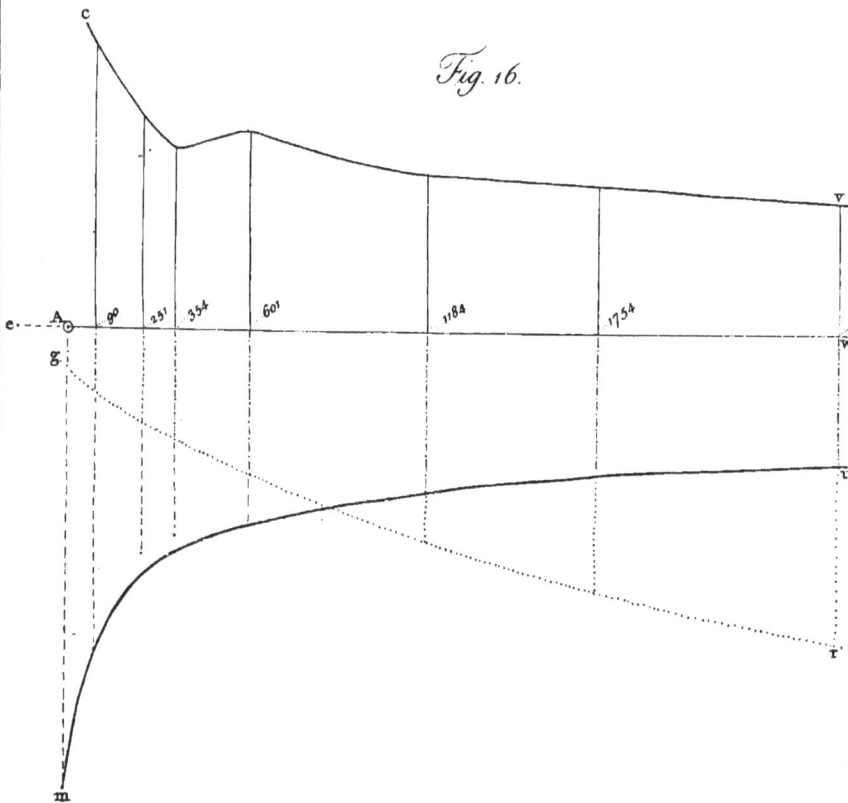

Fig. 16.

c

v

e A 90 231 334 601 1184 1754 w

g

i

r

m

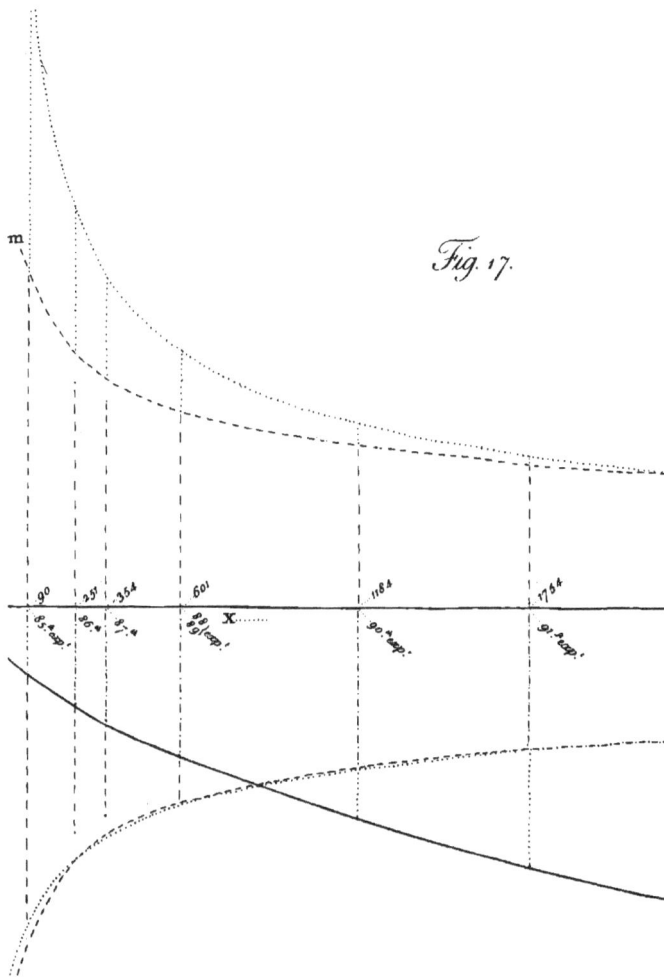

Fig. 17.

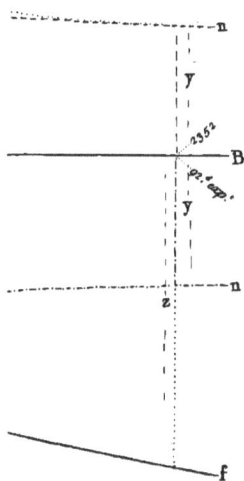

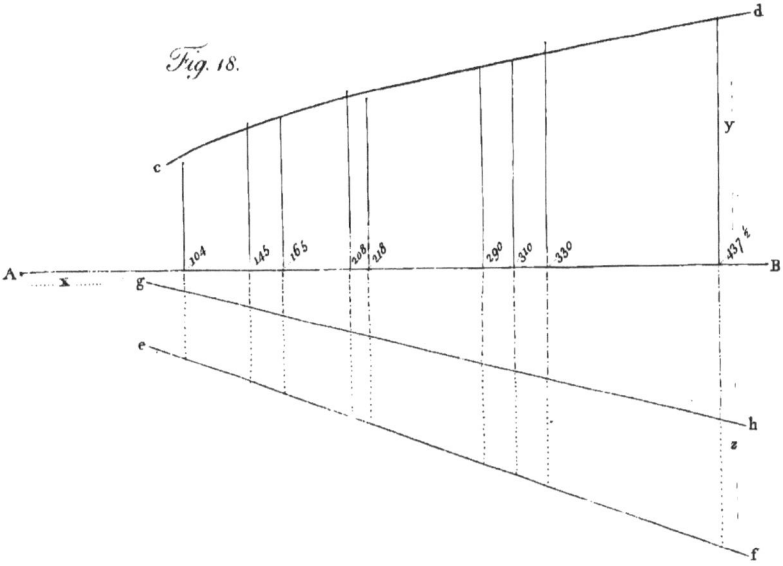

Fig. 18.

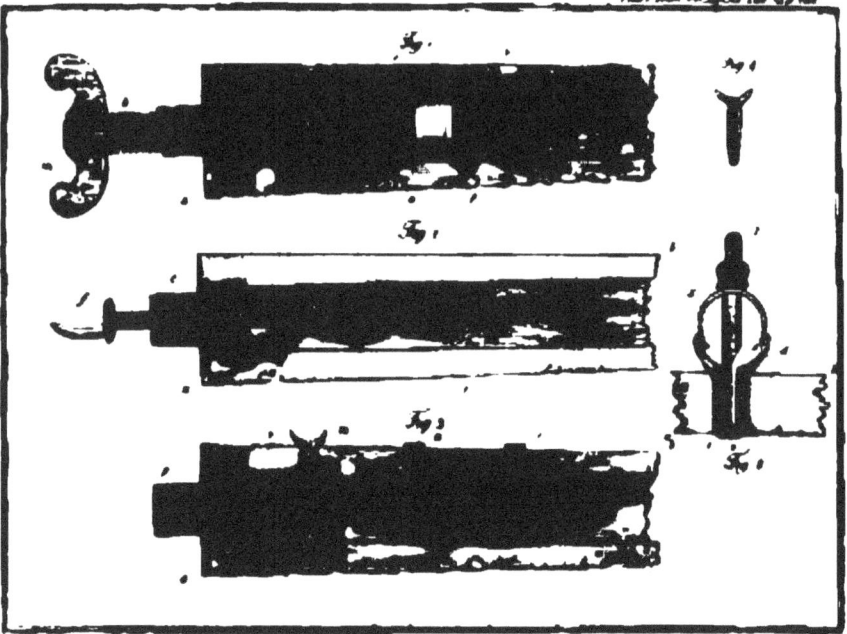